T0135376

Studies in Computational Intelligence

Volume 668

Series editor

Janusz Kacprzyk, Polish Academy of Sciences, Warsaw, Poland
e-mail: kacprzyk@ibspan.waw.pl

About this Series

The series "Studies in Computational Intelligence" (SCI) publishes new developments and advances in the various areas of computational intelligence—quickly and with a high quality. The intent is to cover the theory, applications, and design methods of computational intelligence, as embedded in the fields of engineering, computer science, physics and life sciences, as well as the methodologies behind them. The series contains monographs, lecture notes and edited volumes in computational intelligence spanning the areas of neural networks, connectionist systems, genetic algorithms, evolutionary computation, artificial intelligence, cellular automata, self-organizing systems, soft computing, fuzzy systems, and hybrid intelligent systems. Of particular value to both the contributors and the readership are the short publication timeframe and the worldwide distribution, which enable both wide and rapid dissemination of research output.

More information about this series at http://www.springer.com/series/7092

Cristian Lai · Alessandro Giuliani
Giovanni Semeraro

Editors

Information Filtering and Retrieval

DART 2014: Revised and Invited Papers

 Springer

Editors
Cristian Lai
Center of Advanced Studies, Research and
    Development in Sardinia (CRS4)
Parco Scientifico e Tecnologico della
    Sardegna
Pula
Italy

Giovanni Semeraro
Department of Computer Science
University of Bari Aldo Moro
Bari
Italy

Alessandro Giuliani
Department of Electrical and Electronic
    Engineering
University of Cagliari
Cagliari
Italy

ISSN 1860-949X   ISSN 1860-9503   (electronic)
Studies in Computational Intelligence
ISBN 978-3-319-83449-8   ISBN 978-3-319-46135-9   (eBook)
DOI 10.1007/978-3-319-46135-9

Printed on acid-free paper

This Springer imprint is published by Springer Nature
The registered company is Springer International Publishing AG
The registered company address is: Gewerbestrasse 11, 6330 Cham, Switzerland

Preface

With the increasing availability of data, it becomes more important to have automatic methods to manage data and retrieve information. Data processing, especially in the era of social media, is changing users' behaviors. Users are ever more interested in information rather than in mere raw data. Considering that the large amount of accessible data sources is growing, novel systems providing effective means of searching and retrieving information are required. Therefore, the fundamental goal is to make information exploitable by both humans and machines.

This volume focuses on new challenges and emerging ideas in distributed information filtering and retrieval. It collects invited chapters and extended research contributions from DART 2014 (the 8th International Workshop on Information Filtering and Retrieval), held in Pisa (Italy), on December 10, 2014, and co-located with the XIII AI*IA Symposium on Artificial Intelligence.

The book is focused on new research challenges in intelligent information filtering and retrieval.

In Chapter "Time Event Extraction to Boost an Information Retrieval System", by Pierpaolo Basile, Annalina Caputo, Giovanni Semeraro, and Lucia Siciliani, an innovative information retrieval system able to manage temporal information is proposed. The system allows temporal constraints in a classical keyword-based search. Information about temporal events is automatically extracted from text at indexing time and stored in an ad hoc data structure exploited by the retrieval module for searching relevant documents.

Chapter "Interactive Text Categorisation: The Geometry of Likelihood Spaces", by Giorgio Maria Di Nunzio, presents a two-dimensional representation of probabilities called likelihood spaces. Particularly, the geometrical properties of Bayes' rule when projected into this two-dimensional space are showed, and this concept is extended to naïve Bayes classifiers. This geometrical interpretation is applied to a real machine learning problem of text categorization and a Web application that implements all the concepts on a standard text categorization benchmark is presented.

Chapter "Mining Movement Data to Extract Personal Points of Interest: A Feature Based Approach", by Marco Pavan, Stefano Mizzaro, and Ivan

Scagnetto, proposes a novel approach able to address the aspect of the identification of important locations, i.e., places where people spend a fair amount of time during their daily activities. The proposed method is organized in two phases: first, a set of candidate stay points is identified by exploiting some state-of-the-art algorithms to filter the GPS-logs; then, the candidate stay points are mapped onto a feature space having as dimensions the area underlying the stay point, its intensity (e.g., the time spent in a location), and its frequency (e.g., the number of total visits).

In Chapter "SABRE: A Sentiment Aspect-Based Retrieval Engine, by Annalina Caputo, Pierpaolo Basile, Marco de Gemmis, Pasquale Lops, Giovanni Semeraro, and Gaetano Rossiello, SABRE, a two-stage sentiment aspect-based retrieval engine is proposed. SABRE retrieves opinions about an entity at two different levels of granularity, called aspect and sub-aspect. Such fine-grained opinion retrieval enables both an aspect-based sentiment classification of text fragments and an aspect-based filtering during the navigational exploration of the retrieved documents.

In Chapter "Monitoring and Supporting People that Need Assistance: The BackHome Experience", by Xavier Rafael-Palou, Eloisa Vargiu, Stefan Dauwalder, and Felip Miralles, a sensor-based telemonitoring system is presented. People that need assistance, e.g., the elderly or disabled people, may be affected by a decline in daily functioning that usually involves the reduction and discontinuity in daily routines and a worsening in the overall quality of life. The proposed system addresses all that issues.

Chapter "The Relevance of Providing Useful and Personalized Information to Therapists and Caregivers in Tele", by Juan Manuel Fernandez, Marc Solà, Alexander Steblin, Eloisa Vargiu, and Felip Miralles, presents a generic Tele* (i.e., telemedicine, telerehabiliation, telemonitoring, telecare, and teleassistance) solution that, in principle, may be customized to whatever kind of real scenarios to give a continuous and efficient support to therapists and caregivers. The aim of the proposed solution is to be as flexible as possible in order to be able to provide telerehabilitation, telemonitoring, teleassistance or a conjunction of them, depending on the real situation. Three customizations of the generic platform are also presented.

The main focus of DART was to discuss and compare suitable novel solutions based on intelligent techniques and applied to real-world applications. The chapters of this book present a comprehensive review of related work and state-of-the-art techniques. The authors, both practitioners and researchers, shared their results in several topics such as "Information Retrieval", "Text Categorization", and "Data Mining".

Pula, Italy   Cristian Lai
Cagliari, Italy   Alessandro Giuliani
Bari, Italy   Giovanni Semeraro
October 2015

Contents

Contents

Time Event Extraction to Boost an Information Retrieval System

Pierpaolo Basile, Annalina Caputo, Giovanni Semeraro
and Lucia Siciliani

Abstract In this chapter we propose an innovative information retrieval system able to manage temporal information. The system allows temporal constraints in a classical keyword-based search. Information about temporal events is automatically extracted from text at indexing time and stored in an ad-hoc data structure exploited by the retrieval module for searching relevant documents. Our system can search textual information that refers to specific period of times. We perform an exploratory case study indexing all Italian Wikipedia articles.

1 Introduction

Identifying specific pieces of information related to a particular time period is a key task for searching past events. Although this task seems to be marginal for Web users [17], many search domains, like enterprise search, or lately developed information access tasks, such as Question Answering [19] and Entity Search, would benefit from techniques able to handle temporal information.

The capability of extracting and representing temporal events mentioned in a text can enable the retrieval of documents relevant for a given topic pertaining to a specific time. Nonetheless, the notion of *temporal* in the retrieval context has often being associated with the dynamic dimension of a piece of information, i.e. how it changes over time, in order to promote freshness in results. Such kind of approaches focus on when the document was published (*timestamp*) rather than the temporal event mentioned in its content (*focus time*). While traditional search engines take

P. Basile (✉) · A. Caputo · G. Semeraro · L. Siciliani
Department of Computer Science, University of Bari Aldo Moro, Bari, Italy
e-mail: Pierpaolo.Basile@uniba.it

A. Caputo
e-mail: Annalina.Caputo@uniba.it

G. Semeraro
e-mail: Giovanni.Semeraro@uniba.it

L. Siciliani
e-mail: siciliani.lu@gmail.com

© Springer International Publishing AG 2017   1
C. Lai et al. (eds.), *Information Filtering and Retrieval*,
Studies in Computational Intelligence 668, DOI 10.1007/978-3-319-46135-9_1

into account temporal information related to a document as a whole, our search engine aims to extract and index single events occurring in the texts, and to enable the retrieval of topics related to specific temporal events mentioned in the documents. In particular, we are interested in retrieving documents that are relevant for the user query, and also match some temporal constraints. For example, the user could be interested in a particular topic—strumenti musicali (*musical instrument*)—related to a specific time period—inventati tra il 1300 ed il 1500 (*invented between 1300 and 1500*).

However, looking for happenings in a specific time span requires further, and more advanced, techniques able to treat temporal information. Therefore, our goal is to merge features of both information retrieval (IRS) and temporal extraction systems (TES). While an IRS allows us to handle and access the information included in texts, TES locate temporal expressions. We define this kind of system "Time-Aware IR" (TAIR).

In the past, several attempts have been made to exploit temporal information in IR systems [2], with an up-to-date literature review and categorization provided in [7]. Most of these approaches exploit time information related to the document in order to improve the ranking (recent documents are more relevant) [9], cluster documents using temporal attributes [1, 3], or exploit temporal information for effectively present documents to the user [16]. However, just a handful of work have focused on temporal queries, that is the capability of querying a collection with both free text and temporal expression [4]. Alonso et al. pointed out as this kind of tasks needs the combination of results from both the traditional keyword-based and the temporal retrieval that can give rise to two different result sets. Vandenbussche and Teissèdre [22] dealt with temporal search in the context of both the Web of Content and the Web of Data, but differently from our system, they relied on an ontology of time for temporal queries [11]. Kanhabua and Nørvåg [13] defined semantic- and temporal-based features for a learning to rank approach by extracting named entities and temporal events from the text. Similarly to our approach, Arikan et al. [5] considered the query as composed by a keyword and a temporal part. Then, the two queries were addressed by computing two different language model-based weights. Exploiting a similar model, Berberich et al. [6] developed a framework for dealing with uncertainty in temporal queries. However, both approaches drawn the probability of the temporal query out of the whole document, thus neglecting the pertinence of temporal events at a sentence level. In order to overcome such a limitation, Matthews et al. [16] introduced two different types of indexes, at a document and a sentence level, with the latter associated with content date.

Preliminary to indexing and retrieval, the information extraction phase aims to extract temporal information, and its associated events, from text. In this area [15], several approaches aim at building structured knowledge sources of temporal events. In [12] the authors describe an extension of the YAGO knowledge base, in which entities, facts, and events are anchored in both time and space. Other work exploit Wikipedia to extract temporal events, such as those reported in [10, 14, 24]. Temporal extraction systems can locate temporal expressions and normalize them making this information available for further processing. Currently, there are different tools that can make this kind of analysis on documents, like SUTime [8] or HeidelTime [20] and other systems which took part in TempEval evaluation campaigns. Temporal extraction is not the main focus of this chapter, then we remand the interested reader to the TempEval description task papers [21, 23] for a wider overview of the latest state-of-the-art temporal extraction systems.

The chapter is organized as follows: Sect. 2 provides details about the model behind our TAIR system, while Sect. 3 describes the implementation of our model. Section 4 reports some use cases of the TAIR system which show the potential of our approach, while Sect. 5 closes the chapter.

2 Time-Aware IR Model

A TAIR model should be able to tackle some problems that emerge from temporal search [22], that is: (1) the extraction and normalization of temporal references, (2) the representation of the temporal expressions associated to documents, and (3) the ranking under the constraint of keyword- and temporal-queries.

Our TAIR model consists of three main components responsible to deal with these issues, as sketched in Fig. 1.

Text processing It automatically extracts time expressions from text. The extracted expressions are normalized in a standard format and sent to the indexing compo-
    nent;
Indexing This component is dedicated to index both textual and temporal infor-
    mation. During the indexing, text fragments are linked to time expressions. The
    idea behind this approach is that the context of a temporal expression is relevant;

Fig. 1 The IR time-aware
model

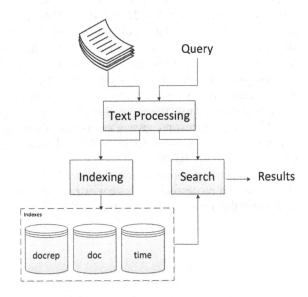

Search It analyzes the user query composed by both keywords and temporal constraints, and performs the search over the index in order to retrieve relevant information.

2.1 Text Processing Component

Given a document as input, the text processing component provides as output the *normalized* temporal expressions extracted from the text, along with information about positions in which the temporal expressions are found. For this purpose we adopt a standard annotation language for temporal expressions called TimeML [18]. We are interested in expressions tagged with the TIMEX3 tag that is used to mark up explicit temporal expressions, such as times, dates and durations. In TIMEX3 the value of the temporal expression is normalized according to 2002 TIDES guideline, an extension of the ISO-8601 standard, and is stored in an attribute called *value*. An example of TIMEX3 annotation for the sentence "before the 23th May 1980" is reported below:

```
<TimeML>
    before the
    <TIMEX3 tid="t3" type="DATE" value="1980-05-23">
        23th May 1980
    </TIMEX3>
</TimeML>
```

Where `tid` is a unique identifier, `type` can assume one of the types between: `DATE`, `TIME`, `DURATION`, and `SET`, while the `value` attribute contains the temporal information that varies accordingly to the type.

ISO-8601 normalizes temporal expressions in several formats. For example, "May 1980" is normalized as "1980–2005", while "23th May 1980" as "1980-05-23". We choose to normalize all dates using the pattern *yyyy-mm-dd*. All temporal expressions not compliant to the pattern, such as "1980", must be normalized retaining the lexicographic order between dates. Our solution consists in normalizing all temporal expressions in the form of *yyyy* or *yyyy-mm* to the last day of the previous year or month, respectively. In our previous example, the expression "1980" is normalized as 19791231. Similarly, the expression "1980–2005" is normalized as "1980-04-30". Moreover, the text processing component applies several normalization rules to correctly identify seasons, for example the TimeML tag for Spring "yyyy-SP" is normalized as "yyyy-03-20".

Using the correct normalization, the order between periods is respected. In conclusion the text processing component extracts temporal information and correctly normalized them to make different time periods comparable.

2.2   The Indexing Component

After the text processing step, we need to store and index data. In our model we propose to store both documents and temporal expressions in three separated data indexes, as reported in Fig. 1.

The first index (*docrep*) stores the text of each document (without processing) with an id, a numeric value that unequivocally identifies the document. This index is used to store the document content only for the presentation purpose. The second index (*doc*) is a traditional inverted index in which the text of each document is indexed and used for keyword-based search. Finally, the last index (*time*) stores temporal expressions found in each document. For each temporal expression, we store the following information:

- The document id.
- The normalized value of the time expression according to the normalization procedure described in Sect. 2.1.
- The start and end offset of the expression in the document, useful for highlighting.
- The context of the expression: the context is defined by taking all the words that can be found within n characters before and after the time expression. The context is indexed and used by the search component during the retrieval step. The idea is to keep trace of the context where the time expression occurred. The context is tokenized and indexed and exploited in conjunction with the keyword-based search, as we explained in Sect. 2.3.

It is important to note that a document could have many temporal expressions, for each of these an entry in the *time* index is created. For example, given the Wikipedia page in Fig. 2, we store its whole content as reported in Table 1a, while we tokenize

Clavicembalo

Da Wikipedia, l'enciclopedia libera.

Con il termine **clavicembalo** (altrimenti detto **gravicembalo**, arpicordo, cimbalo, cembalo) si indica una famiglia di strumenti musicali a corde, dotati di tastiera: tra questi, anzitutto lo strumento di grandi dimensioni attualmente chiamato clavicembalo, ma anche i più piccoli virginale e spinetta.

Questi strumenti generano il suono pizzicando la corda, anziché colpirla come avviene nel pianoforte o nel clavicordo. La famiglia del clavicembalo ha probabilmente avuto origine quando una tastiera è stata adattata ad un salterio, fornendo così un mezzo per pizzicare le corde. Il termine stesso, che compare per la prima volta in un documento del 1397[1], deriva dal latino *clavis*, chiave (intesa come il meccanismo che utilizza il movimento del tasto per azionare il leveraggio retrostante), e *cymbalum*, termine che designava nel medioevo gli strumenti musicali con corde parallele tese su una cassa poligonale e senza manico, come i salteri e le cetre. In ogni caso, la più antica descrizione nota del clavicembalo risale al 1440 circa[2]. I costruttori di clavicembali e strumenti simili sono detti **cembalari** o **cembalai**[3].

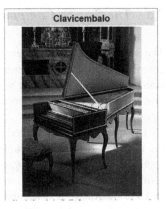
Clavicembalo

Fig. 2 Wikipedia page example

Table 1 The three indices used by the system

Field	Value	Field	Value
(a) docrep intex		*(b) doc index*	
ID	42	ID	42
Content	Con il termine clavicembalo (altrimenti detto gravicembalo, arpicordo, cimbalo, cembalo) si indica una famiglia di strumenti musicali a corde [...]	Content	{ 'Con', 'il', 'termine', 'clavicembalo', 'altrimenti', 'detto', 'gravicembalo', 'arpicordo', 'cimbalo', 'cembalo', 'si', 'indica', 'una', 'famiglia', 'di', 'strumenti','musicali', 'a', 'corde' [...] }
Field	Value		
(c) time index			
ID	42		
Time	13961231		
Start Offset	350		
End Offset	354		
Context	{ '*Il*', '*termine*', '*stesso*', '*che*', '*compare*', '*per*', '*la*', '*prima*', '*volta*', '*in*', '*un*', '*documento*', '*del*', **'deriva'**, **'dal'**, **'latino'**, **'clavis'**, **'chiave'** [...] }		

and index the page as shown in Table 1b. The most interesting part of the indexing step is the storage of temporal expressions. As depicted in Table 1c, for each temporal expression we store the normalized time value, in this case "13961231", and the start and end offset of the expression in the text. Finally, we tokenize and index the context in which the expression occurs. In Table 1c, in italics is reported the left context, while the right context is reported in bold. Examples are reported according to the Italian version of Wikipedia, but the indexing step is language independent.

2.3   The Search Component

The search component retrieves relevant documents according to the user query q containing temporal constraints. For this reason we need to make temporal expressions in the query compliant with the expressions stored in the index. The query is processed by the text component in order to extract and normalize the time expressions.

The query q is represented by two parts: q_k contains keywords, while q_t only the normalized time expressions. q_k is used to retrieve from the *doc* index a first results set RS_{doc}. Thus, both q_k and q_t are used to query the *time* index producing the results set RS_{time}. The search in *time* index is limited to those documents belonging to RS_{doc}. In RS_{time}, text fragments have to match the time constraints expressed in q_t, while the matching with the keyword-based query q_k is optional. The optional matching with q_k has the effect of promoting those contexts that satisfy both the temporal constraints and the query topics, while not completely removing poorly matching results. The motivation behind this approach is twofold: through RS_{doc} we retrieve those documents relevant for the query topic, while RS_{time} contains the text fragments that match the time query q_t and are related to the query topic.

For example given the query q = "clavicembalo [1300 TO 1400]", we identify the two fields: q_k = "clavicembalo" and q_t = [12991231 TO 13991231]. It is important to underline that in this example we adopted a particular syntax to identify range queries, more details about the system implementation are reported in Sect. 3.

The retrieval step produces two results sets: RS_{doc} and RS_{time}. Considering the query q in the previous example: RS_{doc} contains the doc 42 with a relevance score s_{doc}. While the results set RS_{time} contains the temporal expression reported in Table 1c with a score s_{time}. The last step is to combine the two results sets. The idea is to promote text fragments in RS_{time} that comes from documents that belong to RS_{doc}. We simply boost the score of each result in RS_{time} multiplying its score by the score assigned to its origin document in RS_{doc}. In our example the temporal expression occurring in RS_{time} obtains a final score computed as: $s_{doc} \times s_{time}$. We have chosen to boost score rather than linearly combine them, in this way we avoid the use of combination parameters.

Finally, we sort the re-ranked RS_{time} and provide it to the user as final result of the search. It is important to underline that our system does not produce a list of document as a classical search engine does, but we provide all the text passages that are both relevant for the query and compliant to temporal constraints.

3 System Implementation

We implemented our TAIR model in a freely available system[1] as an open-source software under the GNU license V.3. The system is developed in JAVA and extends the indexing and search open-source API Apache Lucene.[2]

The text processing component is based on the HeidelTime-1.8 tool[3] [20] to extract temporal information. We adopt this tool for two reasons: (1) it obtained good performance in the TempEval-3 task, and (2) it is able to analyze text written in several languages including the Italian. HeidelTime is a rule based system that can be extended to support other languages or specific domains.

Our system provides all the expected functionalities: text analysis, indexing and search. The query language supports all operators provided by the Lucene query syntax.[4] Moreover the temporal query q_t can be formulated using natural time expressions, for example "12 May 2014" or "yesterday". The search component tries to

---

[1] https://github.com/pippokill/TAIR.

[2] http://lucene.apache.org/.

[3] https://code.google.com/p/heideltime/.

[4] http://lucene.apache.org/core/4_8_1/queryparser/org/apache/lucene/queryparser/classic/package-summary.html.

Table 2 Example of time query operators

Query	Description
20020101	Match exactly 1st January 2002
[20020101 TO 20030101]	Match from 1st January 2002 to 1st January 2003
[* TO 20030101]	Before 1st January 2003
[20020101 TO *]	After 1st January 2002
01??2002	Any first day of the month in 2002, * should be used for multiple character match, for example 01*2002
20020101 AND 20020131	The first and last day of January 2002, AND and OR operator can be used to combine exact match and range query

automatically translate the user query in the proper time expressions. However, the user can directly formulate q_t using normalized time expressions and query operators. Table 2 shows some time operators.

Currently the system does not provide a GUI for searching and visualizing the results, but it is designed as an API. As future works we plan to extend the API with REST Web functionalities.

4 Use Case

We decided to set up a case study to show the potentialities of the proposed IR framework. The case study involves the indexing of a large collection of documents and a set of example queries exploiting specific scenarios in which temporal expressions play a key role. Moreover, another goal is to provide performance information about the system in terms of indexing and query time, and index space.

We propose an exploratory use case indexing all Italian Wikipedia articles. Our choice is based on the fact that Wikipedia is freely available and contains millions of documents with many temporal events. We need to set some parameters: we index only documents with at least 4,000 characters, remove special pages (e.g. category pages), we set the context size in temporal index to 256 characters.

We perform the experiment on a virtual machine with four virtual cores and 32 GB of RAM. Table 3 reports some statistics related to the indexing step. The indexing time is very high due to the complexity of the temporal extraction algorithm and the

Table 3 Indexing performance

Statistics	Value
Number of documents	168,845
Number of temporal expressions	6,615,430
Indexing time (h)	68
Indexing time (doc./min.)	41, 38

Table 4 Results for the query "19810429"

Result rank	Wikipedia page	Time context
1	Paul Breitner	nel **1981**, richiamato da Jupp Derwall, nel frattempo divenuto nuovo commissario tecnico della Germania Ovest, e con il quale aveva comunque avuto accese discussioni a distanza. Il "nuovo debutto" avviene ad Amburgo il **29 aprile** contro l'Austria
2	...E tu vivrai nel terrore! L'aldilà	Warbeck e Catriona McColl, presente nei contenuti speciali del DVD edito dalla NoShame. Accoglienza. Il film uscì in Italia il **29 aprile 1981** e incassò in totale 747.615.662 lire. Distribuito per i mercati esteri dalla VIP International, ottenne un ottimo successo
3	RCS Media Group	L'operazione venne perfezionata il **29 aprile 1981**. Quel giorno una società dell'Ambrosiano (quindi di Calvi), la "Centrale Finanziaria S.p.A." effettuò l'acquisto del 40% di azioni Rizzoli

huge number of documents. We speed up the temporal event extraction implementing a multi threads architecture, in particular in this evaluation we enable four threads for the extraction.

One of the most appropriate scenarios consists in finding events that happened in a specific date. For example, one query could be interested in listing all events happened on 29 April 1981. In this case the time query is "19810429" while the keyword query is empty. The first three results are shown in Table 4.

We report in bold the temporal expressions that match the query. It is important to note that in the first result the year "1981" appears distant from both the month and the day, but the Text Processing component is able to correctly recognize and normalize the date.

Another interesting scenario is to find events related to a specific topic in a particular time period. For example, Table 5 reports the first three results for the query: "terremoti tra il 1600 ed il 1700" (*earthquakes between 1600 and 1700*). This query is split in its keyword q_k ="terremoti" (*earthquakes*) and temporal component $q_t = [15991231 \text{ TO } 16991231]$.

Table 6 shows the usage of time query operators, in particular of wild-cards. We are interested in facts related to *computers* which happened in January 1984 using the time query pattern "198401??".

As reported in Table 6, the first two results regard events whose time interval encompasses the time expressed in the query, since they took place in 1984, while the third result shows an event that completely fulfil the time requirements expressed in the temporal query.

Table 5 Results for the query "earthquakes between 1600 and 1700"

Result rank	Wikipedia page	Time context
1	Terremoto della Calabria dell'8 giugno 1638	Il terremoto dell'**8 giugno 1638** fu un disastroso terremoto che colpì la Calabria, in particolare il Crotonese e parte del territorio già colpito nei giorni **27 e 28 marzo del 1638**
2	Eruzione dell'Etna del 1669	**1669 10 marzo** − M = 4.8 Nicolosi Terremoto con effetti distruttivi nel catanese in particolare a Nicolosi in seguito all'eruzione dell'Etna conosciuta come Eruzione dell'Etna del **1669**. Il **25 febbraio** e l'**8 e 10 marzo del 1669** una serie di violenti terremoti
3	Terremoto del Val di Noto del 1693	l'evento catastrofico di maggiori dimensioni che abbia colpito la Sicilia orientale in tempi storici.Il terremoto del **9 Gennaio 1693**

Table 6 Results for the query "computer" with the temporal pattern "198401??"

Result rank	Wikipedia page	Time context
1	Apple III	L'Apple III, detto anche Apple ///, fu un personal computer prodotto e commercializzato da Apple Computer dal 1980 al **1984** come successore dell'Apple II
2	Home computer	Apple Macintosh (**1984**), il primo home/personal computer basato su una interfaccia grafica, nonch il primo a 16/32-bit
3	Apple Macintosh	Apple Computer (oggi Apple Inc.). Commercializzato dal **24 gennaio 1984** al 1 ottobre 1985, il Macintosh il capostipite dell'omonima famiglia

Table 7 Results for the query "nato" (born) with the time constraint "[primavera 1980 TO autunno 1980]" ([spring 1980 TO autumn 1980])

Result rank	Wikipedia page	Time context
1	Binningen (Svizzera)	Gianluca Bazzoli, (nato il **2 maggio 1980**), attore
2	SunSet Swish	conosciuti anche con l'acronimo SSS - sono un gruppo musicale J-Pop giapponese. Membri - Daisuke Saeki nato il **28 agosto 1980**. Cantante. - Yki Tomita nato il **13 luglio 1980**. Chitarrista. - Junz? Ishida nato il 28 febbraio 1981. Pianista
3	Foggia	pugile Giuseppe Colucci (nato a Foggia il **24 agosto 1980**)

Table 7 reports results about time constraints expressed in written form, for example "[primavera 1980 TO autunno 1980]" ([spring 1980 TO autumn 1980]). In this case the keyword query is *nato (born)*.

5   Conclusions and Future Work

We proposed a "Time-Aware" IR system able to extract, index, and retrieve temporal information. The system expands a classical keyword-based search through temporal constraints. Temporal expressions, automatically extracted from documents, are indexed through a structure that enables both keyword- and time-matching. As a result, TAIR retrieves a list of text fragments that match the temporal constraints, and are relevant for the query topic. We proposed a preliminary case study indexing all the Italian Wikipedia and described some retrieval scenarios which would benefit from the proposed IR model.

As future work we plan to improve both recognition and normalization of time expressions, extending some particular TimeML specifications that in this preliminary work were not taken into account during the normalization process. Moreover, we will perform a deep "in-vitro" evaluation on a standard document collection.

Acknowledgments The computational work has been executed on the IT resources made available by two projects financed by the MIUR (Italian Ministry for Education, University and Research) in the "PON Ricerca e Competitività 2007–2013" Program: ReCaS (Azione I—Interventi di rafforzamento strutturale, PONa3_00052, Avviso 254/Ric) and PRISMA (Asse II—Sostegno all'innovazione, PON04a2_A).

References

1. Alonso, O., Gertz, M.: Clustering of search results using temporal attributes. In: Proceedings of the 29th Annual International ACM SIGIR Conference on Research and Development in Information Retrieval, pp. 597–598. ACM (2006)
2. Alonso, O., Gertz, M., Baeza-Yates, R.: On the value of temporal information in information retrieval. SIGIR Forum **41**(2), 35–41 (2007)
3. Alonso, O., Gertz, M., Baeza-Yates, R.: Clustering and exploring search results using timeline constructions. In: Proceedings of the 18th ACM Conference on Information and Knowledge Management, CIKM '09, pp. 97–106. ACM (2009)
4. Alonso, O., Strötgen, J., Baeza-Yates, R.A., Gertz, M.: Temporal information retrieval: challenges and opportunities. In: Proceedings of the 1st International Temporal Web Analytics Workshop (TWAW 2011), vol. 11, pp. 1–8 (2011)
5. Arikan, I., Bedathur, S.J., Berberich, K.: Time will tell: leveraging temporal expressions in IR. In: Baeza-Yates, R.A., Boldi, P., Ribeiro-Neto, B.A., Cambazoglu, B.B. (eds.) Proceedings of the 2ND International Conference on Web Search and Web Data Mining, WSDM 2009, Barcelona, Spain, February 9–11, 2009. ACM (2009)
6. Berberich, K., Bedathur, S., Alonso, O., Weikum, G.: A language modeling approach for temporal information needs. In: Proceedings of the 32nd European Conference on Advances in Information Retrieval, ECIR'2010, pp. 13–25. Springer (2010)

7. Campos, R., Dias, G., Jorge, A.M., Jatowt, A.: Survey of temporal information retrieval and related applications. ACM Comput. Surv. **47**(2), 15:1–15:41 (2014)
8. Chang, A.X., Manning, C.D.: SUTime: a library for recognizing and normalizing time expressions. In: LREC, pp. 3735–3740 (2012)
9. Elsas, J.L., Dumais, S.T.: Leveraging temporal dynamics of document content in relevance ranking. In: Proceedings of the 3rd ACM International Conference on Web Search and Data Mining, WSDM '10, pp. 1–10. ACM (2010)
10. Hienert, D., Luciano, F.: Extraction of historical events from Wikipedia. In: Proceedings of the First International Workshop on Knowledge Discovery and Data Mining Meets Linked Open Data, pp. 25–36 (2011)
11. Hobbs, J.R., Pan, F.: An ontology of time for the semantic web. ACM Trans. Asian Lang. Inf. Process. (Special Issue on Temporal Information Processing) **3**(1), 66–85 (2004)
12. Hoffart, J., Suchanek, F.M., Berberich, K., Weikum, G.: YAGO2: a spatially and temporally enhanced knowledge base from Wikipedia. Artif. Intell. **194**, 28–61 (2013)
13. Kanhabua, N., Nørvåg, K.: Learning to rank search results for time-sensitive queries. In: Proceedings of the 21st ACM International Conference on Information and Knowledge Management, CIKM '12, pp. 2463–2466. ACM (2012)
14. Kuzey, E., Weikum, G.: Extraction of temporal facts and events from Wikipedia. In: Proceedings of the 2nd Temporal Web Analytics Workshop, pp. 25–32. ACM (2012)
15. Ling, X., Weld, D.S.: Temporal information extraction. In: Proceedings of the 24th Conference on Artificial Intelligence (AAAI 2010). Atlanta, GA (2010)
16. Matthews, M., Tolchinsky, P., Blanco, R., Atserias, J., Mika, P., Zaragoza, H.: Searching through time in the New York Times. In: Proceedings of the Fourth Workshop on Human-Computer Interaction and Information Retrieval (HCIR 10), pp. 41–44 (2010)
17. Nunes, S., Ribeiro, C., David, G.: Use of temporal expressions in web search. In: Proceedings of the IR Research, 30th European Conference on Advances in Information Retrieval, ECIR'08, pp. 580–584. Springer (2008)
18. Pustejovsky, J., Castano, J.M., Ingria, R., Sauri, R., Gaizauskas, R.J., Setzer, A., Katz, G., Radev, D.R.: TimeML: robust specification of event and temporal expressions in text. New Dir. Quest. Answ. **3**, 28–34 (2003)
19. Saurí, R., Knippen, R., Verhagen, M., Pustejovsky, J.: Evita: A robust event recognizer for QA systems. In: Proceedings of the Conference on Human Language Technology and Empirical Methods in Natural Language Processing, pp. 700–707. ACL (2005)
20. Strötgen, J., Zell, J., Gertz, M.: HeidelTime: tuning english and developing Spanish resources for TempEval-3. In: 2nd Joint Conference on Lexical and Computational Semantics (*SEM), Volume 2: Proceedings of the 7th International Workshop on Semantic Evaluation, pp. 15–19. ACL (2013)
21. UzZaman, N., Llorens, H., Derczynski, L., Allen, J., Verhagen, M., Pustejovsky, J.: Semeval-2013 task 1: Tempeval-3: Evaluating time expressions, events, and temporal relations. In: 2nd Joint Conference on Lexical and Computational Semantics (*SEM), Volume 2: Proceedings of the 7th International Workshop on Semantic Evaluation, pp. 1–9. ACL (2013)
22. Vandenbussche, P.Y., Teissèdre, C.: Events retrieval using enhanced semantic web knowledge. In: Workshop DeRIVE 2011 (Detection, Representation, and Exploitation of Events in the Semantic Web) in cunjunction with 10th International Semantic Web Conference 2011 (ISWC 2011) (2011)
23. Verhagen, M., Sauri, R., Caselli, T., Pustejovsky, J.: SemEval-2010 Task 13: TempEval-2. In: Proceedings of the 5th International Workshop on Semantic Evaluation, pp. 57–62. ACL (2010)
24. Whiting, S., Jose, J., Alonso, O.: Wikipedia as a time machine. In: Proceedings of the Companion Publication of the 23rd International Conference on World Wide Web Companion, pp. 857–862. International World Wide Web Conferences Steering Committee (2014)

Interactive Text Categorisation:
The Geometry of Likelihood Spaces

Giorgio Maria Di Nunzio

Abstract In this chapter we present a two-dimensional representation of probabilities called likelihood spaces. In particular, we show the geometrical properties of Bayes' rule when projected into this two-dimensional space and extend this concept to Naïve Bayes classifiers. We apply this geometrical interpretation to a real machine learning problem of text categorisation and present a Web application that implements all the concepts on a standard text categorisation benchmark.

1 Introduction

Classification is the task of learning a function that assigns a new unseen object to one or more predefined classes based on the features of the object [26, Chap. 4]. Among the many different approaches presented in the literature, Naïve Bayes (NB) classifiers have been widely recognised as a good trade-off between efficiency and efficacy since they are easy to train and achieve satisfactory results [18]. A NB classifier is a type of probabilistic classifier that uses Bayes' rule to predict the class of the unknown object, and it is based on the simplifying assumption that all the features of the object are conditionally independent given the class. Despite being comparable to other learning methods, these classifiers are rarely among the top performers when trained with default parameters [5]. Indeed, the optimisation of the parameters of NB classifiers is often not adequate, if not missing at all. The usual approach is to set default smoothing constants to avoid arithmetic anomalies given by zero probabilities [29]. Moreover, a probabilistic classifier could be greatly improved by taking into account misclassification costs [14]. The choice of these costs is not trivial and, as for the case of probability smoothing, default costs are used.

By involving users directly in the process of building a probabilistic model, as suggested by [3], one can obtain a twofold result: first, the pattern recognition capabilities of the human can be used to increase the effectiveness of the classifier construction

G.M. Di Nunzio (✉)
Department of Information Engineering, University of Padua,
Via Gradenigo 6/a, 35131 Padua, Italy
e-mail: giorgiomaria.dinunzio@unipd.it

© Springer International Publishing AG 2017   13
C. Lai et al. (eds.), *Information Filtering and Retrieval*,
Studies in Computational Intelligence 668, DOI 10.1007/978-3-319-46135-9_2

and understand why some parameters work better than others; second, visualisation of the model can be used to teach non-experts how probabilistic models work and improve the overall effectiveness of the classification task. Interactive machine learning is a relatively new area of machine learning where model updates are faster and more focused with respect to classical machine learning algorithms; moreover, the magnitude of the update is small; hence, the model does not change drastically with a single update. As a result, even non-expert users can solve machine learning problems through low-cost trial and error or focused experimentation with inputs and outputs. In this respect, the importance of the design of proper user interfaces for the interaction with machine learning models is crucial. Recently, an approach named "Explanatory Debugging" has been described and tested to help end users build useful mental models of a machine learning system while simultaneously allowing them to explain corrections back to the system [17]. The authors found a significant correlation between how participants understood how the learning system operated and the performance of participants' classifiers.

Based on the idea of likelihood spaces [24], we present the geometric properties of the two-dimensional representation of probabilities [7, 8] which allows us to provide an adequate data and knowledge visualisation for understanding how parameter optimisation and cost sensitive learning affect the performance of probabilistic classifiers in a real machine learning setting. We apply this geometrical interpretation to the problem of text categorisation [21], in particular to a standard collection of newswires, the Reuters-21578 collection.[1]

The main objectives of this chapter are:

- A geometrical definition of the Bayes' rule and a discussion on the implications of the normalisation of posterior probabilities.
- An alternative derivation of the likelihood space from the definition of the logit function.
- A description of the link between Bayesian Decision Theory and Likelihood spaces.
- A geometrical definition of NB classifiers.
- An interactive Web application to show how these concepts work in practice both on a toy-problem and on a real case scenario.

This chapter is organized as follows: In Sect. 2, we describe the mathematical background behind the idea of the two dimensional representation. In Sect. 3, we present the details of the likelihood space applied to the NB classifier, in particular the multivariate Bernoulli model. Section 4 is dedicated to the interactive text categorization application on a real machine learning problem. In Sect. 5, we give our final remarks and discuss future works and open research questions.

---

[1]http://www.daviddlewis.com/resources/testcollections/.

1.1 Related Works

The term "interactive machine learning" was probably coined around the very end of the 1990s. A work that paved the way for this research area was a paper on interactive decision tree construction by Ankerst et al. [2]. The same authors also redefined the paradigm that "the user is the supervisor" in this cooperation between humans and machine learning algorithms, that is the system supports the user and the user always has the final decision [3]. In the same years, Ware et al. demonstrated that even users who are not domain experts can often construct good classifiers using a simple two-dimensional visual interface, without any help from a learning algorithm [27]. Ben Shneiderman (author of the "eight golden rules for user interfaces" [22]) gives his impressions on the importance of the effective combination of information visualisation approaches and data mining algorithms in [23]. The first paper that used "interactive machine learning" in the title was by Fails and Olsen [15] in which the authors describe the difference between a classical and an interactive machine learning approach and show an interactive feature selection tool for image recognition. From the point of view of machine learning/artificial intelligence, an excellent survey on the methods and approaches used in the last 15 years has been presented by Amershi et al. [1].

Information visualisation is an important part of the research area of interactive machine learning, in particular for the parts relative to the design of appropriate user interfaces and the possible visualisation choices for classification tasks. For example, in [4], the authors present a framework for a feedback-driven view exploration, inspired by relevance feedback approaches used in Information Retrieval, that makes the exploration of large multidimensional datasets possible by means of visual classifiers. Although we focus less on this part in this paper, we suggest to refer to [9] for a survey on visual classification approaches and to [16] for a survey on text visualisation techniques.[2]

2 Mathematical Background

We suppose to work with a set of n classes $C = \{c_1, \ldots, c_i, \ldots, c_n\}$, and that an object can be assigned to (and may actually belong to) more than one class; this is also known as the problem of overlapping categories. Instead of building one single multi-class classifier, we split this multi-class categorisation into n binary problems; therefore, we have n binary classifiers [21]. A binary classification problem is a special case of single-labels classification task in which each object belongs to one category or its complement. The usual notation to indicate these two classes is: c_i for the 'positive' class and $\bar{c}_i$ for the 'negative' class (we drop the index i and use c and $\bar{c}$ as long as there is no risk of misinterpreting the meaning).

---

[2]http://textvis.lnu.se.

In this first part, we start building a probabilistic classifier which, given an object o and a category $c \in C$, classifies o in c if the following statement is true:

$$P(c|o) > P(\bar{c}|o) \tag{1}$$

that is, if the probability of the class c is greater than the probability of its complement $\bar{c}$ given the object o.

2.1 The Geometry of Bayes' Rule

Bayes' rule gives a simple yet powerful link between prior and posterior probabilities of events. For example, assume that we have two classes c and $\bar{c}$ and we want to classify objects according to some measurable features. The probability that an object o belongs to c, $P(c|o)$, can be computed in the following way[3]:

$$\underbrace{P(c|o)}_{posterior} = \frac{\overbrace{P(o|c)}^{likelihood}\ \overbrace{P(c)}^{prior}}{\underbrace{P(o)}_{evidence}} \tag{2}$$

Bayes' rule tells how, by starting from a prior probability on the category c, $P(c)$, we can update our belief on that category based on the likelihood of the object, $P(o|c)$, and obtain the so-called posterior probability $P(c|o)$. $P(o)$ is the probability of the object o, also known as the evidence of the data. The probability of the complementary category $\bar{c}$ is computed accordingly:

$$P(\bar{c}|o) = \frac{P(o|\bar{c})P(\bar{c})}{P(o)} \tag{3}$$

In the two-dimensional view of probabilities, we can imagine the posterior probabilities as the two coordinates of the object o in a Cartesian space, where $x = P(c|o)$ and $y = P(\bar{c}|o)$. Since the two classes are complementary, the two conditional probabilities sum to one, therefore:

$$P(\bar{c}|o) = 1 - P(c|o), \quad \text{or} \tag{4}$$
$$y = 1 - x \tag{5}$$

---

[3] We are intentionally simplifying the notation in order to have a cleaner description. In particular, when we write $P(c|o)$, we actually mean $P(C = c|O = o)$, where C and O are two random variables, and c and o two possible values, respectively.

Fig. 1 Bayes' rule on a two-dimensional space. The probability of one class is complementary to the other, $P(\bar{c}|o) = 1 - P(c|o)$

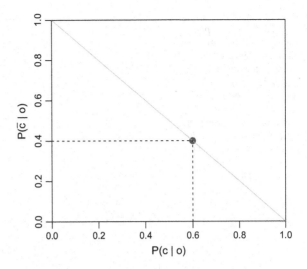

which means that the point with coordinates (x, y) lies on the segment with endpoints $(1, 0)$, $(0, 1)$ in the two dimensional space, as shown in Fig. 1. When we want to classify the object, we compare the two probabilities as already shown in Eq. (1). When we use Bayes' rule to calculate the posterior probabilities, we obtain:

$$\frac{P(o|c)P(c)}{P(o)} > \frac{P(o|\bar{c})P(\bar{c})}{P(o)} \tag{6}$$

It can be immediately seen that we assign o to class c when the probability $P(c|o)$ is greater than 0.5. Since $P(o)$ appears in both sides of the inequality, we can cancel it without changing the result of the classification:

$$P(o|c)P(c) > P(o|\bar{c})P(\bar{c}) \tag{7}$$

remembering that $P(o|c)P(c) \neq 1 - P(o|\bar{c})P(\bar{c})$ since we removed the normalisation factor. An alternative way to cancel $P(o)$ is considering the problem of classification in terms of the odds of the probability $P(c|o)$:

$$P(c|o) > P(\bar{c}|o) \tag{8}$$

$$\frac{P(c|o)}{P(\bar{c}|o)} > 1 \tag{9}$$

$$\frac{P(o|c)P(c)}{P(o|\bar{c})P(\bar{c})} > 1 \tag{10}$$

$$P(o|c)P(c) > P(o|\bar{c})P(\bar{c}) \tag{11}$$

Fig. 2 Bayes' rule without normalisation. The *point* moves from the segment with endpoints $(0, 1)$, $(1, 0)$ towards the origin. In this example, $P(o) = 0.3$. The ratio of the coordinates remain the same as well as the relative position with respect to the *decision line* with angular coefficient $m = 1$ (bisecting line of the first quadrant)

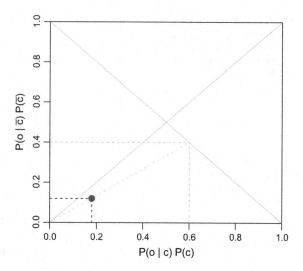

In geometrical terms, the new coordinates x' and y' of the point of the object o are:

$$x' = x P(o) = P(o|c)P(c) \tag{12}$$
$$y' = y P(o) = P(o|\bar{c})P(\bar{c}) \tag{13}$$

The new coordinates are the old ones multiplied by $P(o)$ which means that we are actually 'pushing' the points towards the origin of the axis along the segment with endpoints $(0, 0)$, $(P(c|o), P(\bar{c}|o))$ since both coordinates are multiplied by the same positive number between 0 and 1, as shown in Fig. 2.

Equation (11) can also be interpreted as a decision line with equation $y' = x'$. A more general classification line takes into account an angular coefficient m

$$mx' > y' \tag{14}$$

This non-negative parameter m comes from the introduction of misclassification costs of a Bayesian Decision Theory approach (see Sect. 2.3). Intuitively, when $m = 1$ we count every misclassification (false positives or false negatives) equally. If $m > 1$, we give more importance to the positive class and we are willing to accept more objects in this class; if $m < 1$, we increase the possibility that a point is above the line and classified under the negative category. An alternative, but equivalent, way of looking at this problem is to compare the value of the odds with a threshold k [6]:

$$\frac{x'}{y'} > \frac{1}{m} \tag{15}$$

$$\frac{P(o|c)P(c)}{P(o|\bar{c})P(\bar{c})} > k \tag{16}$$

where k (inversely proportional to m in this formulation) can be set to optimise classification, and it is usually tuned to compensate for the unbalanced classes situation, that is when one of the two classes is much more frequent than the other [19]. This is often the case for any multi-class problem, since the complementary category $\bar{c}$ is about $n - 1$ times bigger than c. This is also the case for real two-class categorisation problems, like spam classification, where the difference in proportion of the number of objects in the two classes 'spam' and 'ham' is very large. We can incorporate this disproportion between the two classes in the angular coefficient m of the two-dimensional space in the following way:

$$mx' > y' \tag{17}$$

$$mP(o|c)P(c) > P(o|\bar{c})P(\bar{c}) \tag{18}$$

$$m\frac{P(c)}{P(\bar{c})}P(o|c) > P(o|\bar{c}) \tag{19}$$

$$m'x'' > y'' \tag{20}$$

where $m' = m\frac{P(c)}{P(\bar{c})}$ is the new angular coefficient of the decision line $y'' = m'x''$, and $x'' = P(o|c)$ and $y'' = P(o|\bar{c})$. At this point we have defined the coordinates of an object in terms of the two likelihood functions $P(o|c)$ and $P(o|\bar{c})$ as shown in Fig. 3.

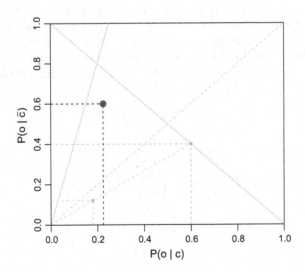

Fig. 3 Data space formed by the coordinates $P(o|c)$ and $P(o|\bar{c})$. This is an example of an unbalanced class situation where the prior $P(c) = 0.7$ is so high that the object is classified under c (in accordance with the earlier examples) despite the likelihood of the object of the negative class is almost three times the one of the positive class. In this example, $m = 1$ and $m' = \frac{P(c)}{P(\bar{c})}$. If we set $m = \frac{P(\bar{c})}{P(c)}$, we would get $m' = 1$ and rebalance the proportion of classes (and change the classification decision). The points of the previous figures are shown in *light grey* for comparison

All the alternatives presented so far are equivalent in terms of classification decisions. There are two connections with two relevant works in the literature that we want to stress: one with the Neyman-Pearson approach [20], and the other with the work of Pazzani and colleagues on the optimality of NB classifiers [12, 28]. The Neyman-Pearson lemma states that the likelihood ratio test defines the most powerful region of acceptance, which is exactly what we have in Eq. (20):

$$\frac{P(o|c)}{P(o|\bar{c})} > M \tag{21}$$

where M is a threshold that defines the region of acceptance. In the optimality of NB classifiers, the authors find an adjustment of the probabilities of the classes $P(c)$ and $P(\bar{c})$ which is again exactly the same idea since we are actually changing the angular coefficient m'.

2.2 Bayes' Rule on Likelihood Space

So far, we have described the two-dimensional representation of the Bayes' rule in the so-called 'data space' which is the space in which the original data resides. The likelihood space, however, is the space formed by the log-likelihood probabilities [24]. The likelihood space can be derived directly by applying the logs of Eq. (20). In this section, we present an alternative way, which is different from the original paper, to obtain the likelihood space which starts from the classification decision given by the log-odds, or logit function, compared to the logarithm of the threshold k:

$$\log\left(\frac{P(c|o)}{P(\bar{c}|o)}\right) > \log(k) \tag{22}$$

$$\log\left(\frac{P(o|c)P(c)}{P(o|\bar{c})P(\bar{c})}\right) > \log(k) \tag{23}$$

$$\log\left(\frac{P(o|c)}{P(o|\bar{c})}\right) + \log\left(\frac{P(c)}{P(\bar{c})}\right) > \log(k) \tag{24}$$

$$\log\left(P(o|c)\right) + \log\left(\frac{P(c)}{P(\bar{c})}\frac{1}{k}\right) > \log\left(P(o|\bar{c})\right) \tag{25}$$

$$x_L + q_L > y_L \tag{26}$$

The likelihood space coordinates of an object o, $x_L = \log\left(x''\right)$ and $y_L = \log\left(y''\right)$, are the logarithms of the coordinates of the data space. An interesting relation between the data space and the likelihood space is that, while in the data space we 'rotate' the decision line around the origin of the axis ($y'' = m'x''$), the same decision line in the likelihood space correspond to a parallel line to the bisecting line of the first and third quadrant $y_L = x_L + q_L$ where $q_L = \log(m')$ is the intercept of this line. In Fig. 4, we show an example of the likelihood space relative to the point of Fig. 3.

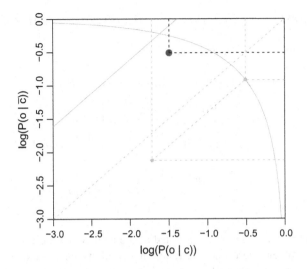

Fig. 4 Bayes' rule on likelihood space. The *red point* corresponds to the one shown in Fig. 3. Note that the *decision line* (*solid grey line*) is *above the red point* as expected. The *decision line* moves parallel to the bisecting line of the third quadrant. In *light grey*, the *points* relative to Figs. 1 and 2. Non normalised points move parallel to the bisecting lines and towards minus infinity, instead of going towards the origin. The segment with endpoints (0, 1), (1, 0) becomes a logarithmic curve in the likelihood space

2.3   Bayesian Decision Theory on Likelihood Spaces

In Bayesian Decision Theory, the objective is to quantify the trade-off between various classification decisions using probabilities and the costs that accompany such decisions [13, Chap. 2]. Whenever we have an object to classify, if we take the decision to classify it under c, we are actually "taking a risk" because we may choose the wrong category. In this framework, the classification of an object becomes the problem of choosing the 'less risky' category; for a binary classification problem, the Bayes decision rule corresponds to selecting the action for which the risk is minimum:

$$R(c|o) < R(\bar{c}|o) \tag{27}$$

$R(c|o)$ and $R(\bar{c}|o)$ are the conditional risks defined as:

$$R(c|o) = \lambda(c|c)P(c|o) + \lambda(c|\bar{c})P(\bar{c}|o) \tag{28}$$

$$R(\bar{c}|o) = \lambda(\bar{c}|c)P(c|o) + \lambda(\bar{c}|\bar{c})P(\bar{c}|o) \tag{29}$$

where $\lambda(\cdot|\cdot)$ is the loss function of an action given the true classification. For example, $\lambda(c|\bar{c})$ quantifies the loss in taking the decision c when the 'true' decision is $\bar{c}$. The new classification decision becomes:

$$\lambda(c|c)P(c|o) + \lambda(c|\bar{c})P(\bar{c}|o) < \lambda(\bar{c}|c)P(c|o) + \lambda(\bar{c}|\bar{c})P(\bar{c}|o) \qquad (30)$$

We can group common terms and obtain:

$$[\lambda(c|\bar{c}) - \lambda(\bar{c}|\bar{c})]P(\bar{c}|o) < [\lambda(\bar{c}|c) - \lambda(c|c)]P(c|o) \qquad (31)$$

$$P(\bar{c}|o) < \frac{[\lambda(\bar{c}|c) - \lambda(c|c)]}{[\lambda(c|\bar{c}) - \lambda(\bar{c}|\bar{c})]}P(c|o) \qquad (32)$$

$$P(o|\bar{c})P(\bar{c}) < \frac{[\lambda(\bar{c}|c) - \lambda(c|c)]}{[\lambda(c|\bar{c}) - \lambda(\bar{c}|\bar{c})]}P(o|c)P(c) \qquad (33)$$

$$P(o|\bar{c}) < \frac{[\lambda(\bar{c}|c) - \lambda(c|c)]}{[\lambda(c|\bar{c}) - \lambda(\bar{c}|\bar{c})]}\frac{P(c)}{P(\bar{c})}P(o|c) \qquad (34)$$

$$y'' < m'x'' \qquad (35)$$

So the ratio of the costs can be interpreted as the angular coefficient m included in m' of Eq. (20). When a zero-one loss function is used, we have $\lambda(c|c) = \lambda(\bar{c}|\bar{c}) = 0$ which means that we have no loss when we give the correct answer, and $\lambda(c|\bar{c}) = \lambda(\bar{c}|c) = 1$ which means that we have a cost equal to one every time we assign the object to the wrong category.

3   Naïve Bayes on Likelihood Space

In real case scenarios, projecting objects into likelihood spaces becomes a necessity since the conditional probabilities $P(o|c)$ and $P(o|\bar{c})$ rapidly go to zero. This problem becomes evident when we use a Naïve Bayes assumption. For example, if o is represented by a set of k features $F = \{f_1, \ldots, f_j, \ldots, f_k\}$, a Naïve Bayes approach allows us to factorize $P(o|c)$ as:

$$P(o|c) = \prod_{j=1}^{k} P(f_j|c) \qquad (36)$$

where features are independent from each other given the class. Suppose that, on average, the probability of a feature given a class is $P(f_j|c) \simeq 10^{-2}$ and all the features have a probability greater than zero to avoid $P(o|c) = 0$. With 100 features, the likelihood of an object will be, on average, $P(o|c) \simeq 10^{-200}$ which is very close to the limit of the representation of a 64 bit floating point number. In real situations, probabilities are much smaller than 10^{-2} and features can be easily tens of thousands; hence, all the likelihood functions would be equal to zero by approximation. Instead, in likelihood spaces, the product becomes a sum of logarithms of probabilities:

Bayesian 2D

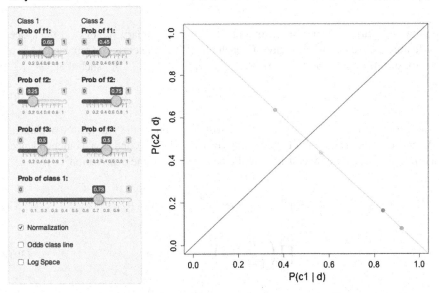

Fig. 5 An interactive demo to show how a multivariate Bernoulli NB model works on a two-dimensional space

$$\log(P(o|c)) = \log \left(\prod_{j=1}^{k} P(f_j|c) \right) = \sum_{j=1}^{k} \log(P(f_j|c)) \qquad (37)$$

In the following section, we derive the mathematical formulation of a NB model that represents features with binary variables, known as multivariate Bernoulli NB model. In Fig. 5, we show a screenshot of an interactive demo of this type of NB classifier.[4] The aim of this toy example is to show the geometric interpretation of this classifier rather than study the optimal parameters for classification. The user can change the conditional probability of each single feature (f_1, f_2, and f_3) and the prior probability of class c_1 (the positive class). The points represent the eight possible combinations (three binary features, hence $2^3 = 8$ objects); when the conditional probability of a feature given the positive class equals that of the negative class, some points overlap in the data space (because we are not able to use that feature to discriminate the objects of one class from the others). The selection widgets allow for choosing normalised probabilities and working in the likelihood space ('log space').

---

[4]http://gmdn.shinyapps.io/bayes2d/.

3.1 Multivariate Bernoulli NB Model

In the multivariate Bernoulli NB model, an object is a binary vector over the space of features. Given a set of features F, each object o of the class c is represented as a vector of k Bernoulli random variables $o \equiv (f_1, \ldots, f_j, \ldots, f_k)$ such that:

$$f_j \sim \text{Bern}\left(\theta_{f_j|c}\right). \tag{38}$$

where $\theta_{f_j|c}$ is the parameter of the Bernoulli distribution for the j-th feature of class c. We can re-write the probability of an object by using the NB conditional independence assumption, this time by considering the parameter θ of the distribution[5]:

$$P(o|c; \theta) = \prod_{j=1}^{k} P(f_j|c; \theta) = \prod_{j=1}^{k} \theta_{f_j|c}^{h_j} \left(1 - \theta_{f_j|c}\right)^{1-h_j}$$

$$= \prod_{j=1}^{k} \left(\frac{\theta_{f_j|c}}{1 - \theta_{f_j|c}}\right)^{h_j} \left(1 - \theta_{f_j|c}\right), \tag{39}$$

where h_j is either 1 or 0 indicating whether feature f_j is present or absent in object o. When we project this probability into the likelihood space, we obtain:

$$\log(P(o|c; \theta)) = \sum_{j=1}^{k} h_j \log \left(\frac{\theta_{f_j|c}}{1 - \theta_{f_j|c}}\right) + \sum_{j=1}^{k} \log(1 - \theta_{f_j|c}), \tag{40}$$

In terms of the likelihood projections, each object of class c has a coordinate composed by: (i) a variable part, the first sum, that depends on the features that are present in the object, and (ii) a fixed part, the second sum, that considers all the features F independently from the features that appear in the object. This second part is very important because, in many works, it is ignored (actually canceled) with the justification that it is a constant independent from the object and, therefore, it does not change the classification decision. This is true only if we do not fix q_L in advance but, on the contrary, we find the optimal parameter q_L of the decision line of Eq. (26). In fact, once q_L is fixed, including or excluding the second sum in the computation of the coordinates would result in a different decision since the points would have different coordinates. The two solutions are equivalent 'only' when we choose an appropriate threshold:

$$\log\left(x''\right) + \log\left(m'\right) > \log(y'') \tag{41}$$

$$\log(x_1'') + \log(x_2'') + \log\left(m'\right) > \log(y_1'') + \log(y_2'') \tag{42}$$

$$\log(x_1'') + \log\left(\frac{m'x_2''}{y_2''}\right) > \log(y_1'') \tag{43}$$

---

[5] We use the notation $P(o|c; \theta)$ to indicate the probability parametrised by θ.

where $x_1'' = \sum_{j=1}^{k} h_j \log \left(\frac{\theta_{f_j|c}}{1-\theta_{f_j|c}} \right)$ and $x_2'' = \sum_{j=1}^{k} \log(1 - \theta_{f_j|c})$ (y_1'' and y_2'' are defined accordingly). For example if we set $m' = 1$ in Eq. (41), then we must set $m' = y_2''/x_2''$ to obtain the same classification in Eq. (43).

3.2 Probability Smoothing

The parameter $\theta_{f|c}$ of each Bernoulli random variable can be estimated in different ways. A common solution is a maximum likelihood approach:

$$\theta_{f|c} = \frac{n_{f,c}}{n_c} \tag{44}$$

where $n_{f,c}$ is the number of objects of category c in which feature f appears, and n_c is the number of objects of category c. However, this approach generates arithmetical anomalies; in particular, a probability equal to zero when the feature is absent in category c, $n_{f,c} = 0$ (or a probability equal to one when $n_{f,c} = n_c$ but it is less frequent). A zero in one of the features of the objects corresponds to a likelihood equal to zero (or a minus infinity in the log space). To avoid these arithmetical problems, smoothing is usually applied. For example, Laplacian smoothing or add-one smoothing:

$$\theta_{f|c} = \frac{n_{f,c} + 1}{n_c + 2} \tag{45}$$

In this chapter, instead of a maximum likelihood approach, we estimate the parameter $\theta_{f|c}$ by using a conjugate prior approach which, in this case, corresponds to finding a *beta* function with parameters α and β [11]:

$$beta_{f|c} = \theta_{f|c}^{\alpha-1}(1 - \theta_{f|c})^{\beta-1}. \tag{46}$$

The result of this choice is that the estimate $\theta_{f|c}$ is now governed by the two hyper-parameters α and β in the following way:

$$\theta_{f|c} = \frac{n_{f,c} + \alpha}{n_c + \alpha + \beta} \tag{47}$$

note that for $\alpha = \beta = 1$, we obtain the Laplacian smoothing. It is possible to optimise α and β for each feature, but in this work we choose to use the same parameters for all the features.

3.3   Decision Line in Likelihood Spaces

As suggested by the authors of the original paper of likelihood spaces [24], one advantage with working in likelihood spaces is that we can devise new strategies for classifying objects. In fact, if we do not limit ourselves to the Bayesian Decision Theory, we can find other linear or non-linear solutions that work much better in terms of classification. The first improvement would be to add a 'rotation' to the decision line in the likelihood space. The authors of the seminal paper discuss this problem and show that polynomial decision lines in the likelihood space can obtain a significant improvement in terms of classification accuracy. However, a polynomial line in the likelihood space corresponds to a complex curve in the data space. Suppose that we find a decision function of this type

$$y_L < m_L x_L + q_L \tag{48}$$

where y_L, x_L, q_L are the same as Eq. (26) and m_L is the angular coefficient of the new decision line. This corresponds to:

$$e^{y_L} < e^{m_L x_L + q_L} \tag{49}$$

$$e^{\log(P(o|\bar{c}))} < e^{m_L \log(P(o|c)) + q_L} \tag{50}$$

$$P(o|\bar{c}) < P(o|c)^{m_L} e^{q_L} \tag{51}$$

which is a sort of exponential curve in the data space. Alternatively, it is also possible to show that a rotation and a shift of the decision function in the data space corresponds to a non-linear curve in the likelihood space [8]. However, it is not our main objective to discuss the possible extension of Bayesian Decision Theory in this chapter. However, we want to stress the fact that, for the interactive text categorisation problem, we use a decision line in the likelihood space like the one shown Eq. (48) but this choice does not have an immediate interpretation in the data space in terms of Bayesian Decision Theory.

4   An Example of Interactive Text Categorization

In the previous sections, we presented the geometric interpretation of probabilistic classifiers on a two-dimensional space, and we described a set of parameters that can be tuned to optimise classification. In particular:

- We can change the estimates of the probability of the features by modifying the values α and β of the prior beta function.
- We can adjust the classification line by changing the intercept q_L and the angular coefficient m_L in the likelihood space.

Table 1 Number of training documents for each class of the Reuters-21578 collection

Category	Training
Acq	1650
Corn	182
Crude	391
Earn	2877
Grain	434
Interest	347
Money-fx	539
Ship	198
Trade	369
Wheat	212
Total	6494

In a real machine learning setting these parameters need to be trained and validated using portions of the dataset available to train the classifier. For example, a k-fold cross validation can be used to find the parameters that minimise the error of the classifier [13, Chap. 9]. For this reason, we have developed an interactive application that allows users to see how the tuning of these parameters affects classification on a real text classification problem.[6]

The top 10 most frequent categories of the Reuters-21578[7] corpus were chosen as a benchmark. In particular, we chose the 6494 training documents. Table 1 shows the number of training documents for each category. Some text preprocessing was done: a first cleaning was done to remove all the punctuation marks and convert all the letters to lowercase. A stoplist of 571 words and contractions (that is, 're, don't, etc.) was used to remove the most frequent words of the English language.[8] Finally, the English Porter stemmer[9] was used as the only method to reduce the space of terms.

Standard classification measures are calculated for the k-fold cross validation and shown real time as parameters are tuned [25].

4.1 Description of the Interface

The main window is split into two parts: the sidebar on the left and the main panel on the right, as shown in Fig. 6. On the left side, the user can interact with the classifier and see the results on the right in terms of both the accuracy of the classification and the visualisation.

---

[6]http://gmdn.shinyapps.io/shinyK.

[7]http://www.daviddlewis.com/resources/testcollections/.

[8]http://jmlr.org/papers/volume5/lewis04a/a11-smart-stop-list/.

[9]http://www.tartarus.org/~martin/PorterStemmer/.

Reuters-21578 Data

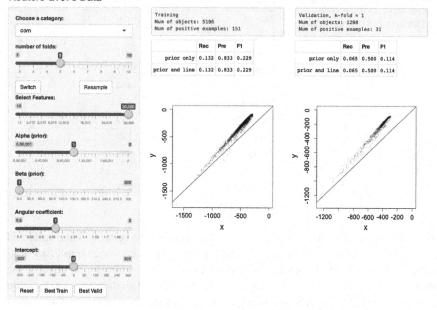

Fig. 6 Interactive text categorisation. Default values of a multivariate Bernoulli NB classifier on the Reuters-21578 dataset

4.1.1 Interaction

The user can interact with the classifier by adjusting and changing the values of the following widgets (we describe them from top to bottom, but the user can interact in any order):

1. The user chooses the category of documents to classify with a selection input menu.
2. The number of k-folds, between 2 and 10, to train and validate on are selected by a slider; the user can also switch from one k-fold to the other (for example, with five folds, the first fold is used for validation while the other four folds are used to train the classifier), or re-sample the folds (documents are randomly sampled to create a new k-fold cross validation) by using the two buttons below the slider of the number of folds.
3. The number of features (terms) can be selected with a slider from 5 up to 30,000 features.
4. The parameters of the beta prior can be adjusted by the two sliders *Alpha*, from 10^{-5} to 2, and *Beta*, from 0.5 to 300.
5. The decision line can be adjusted with the two sliders *Angular coefficient*, values from 0.5 to 2, and *Intercept*, values in the range $-300, 300$.

6. The user can reset all the parameters to the default values, or go back to the best settings found for the training set or the validation set by using one of the three buttons.

4.1.2 Visualization

The main panel is divided into two columns: the first column shows the results on the training set, the second column the results on the validation set. Both columns contain the following information (from top to bottom):

1. The text box shows the total number of objects and the number of positive examples (red points, the documents of the chosen category). The box in the validation column also tells the user on what fold we are validating.
2. The table shows performance measures in terms of Recall, Precision and F1. The first row displays the performance of the classifier when only the parameter of the priors are used, while the second row gives the results when both the prior and the coefficient of the decision line are taken into account.
3. The two-dimensional plot shows in red the documents of the chosen class and in black all the other documents of the collection. The blue line changes according to the parameters *Angular coefficient* and *Intercept*, m and q respectively, while the green line (visible only when the previous parameters are not the default ones) remains fixed to the bisecting line of the third quadrant.

4.2 Example of Usage

Figure 6 shows an example of one category '*corn*' that is quite unbalanced, since the number of positive examples of this category is around 180 and the total number of training examples is about 6400. In order to recover this disproportion, we can change the value of the intercept of the decision line and increase it to 200. In this way, we get an almost perfect recall but the precision is low, as shown in Fig. 7. This situation shows how the intervention of the loss function (which influences the shift of the line in the likelihood space) is good but not optimal. A rotation of the line can significantly improve the situation as shown in Fig. 8.

This optimisation can continue iteratively by slightly changing the intercept and the angular coefficient. Additionally (or alternatively), the user can change the smoothing of the probabilities with the sliders alpha and beta. As surprising as it may seem, for small values of alpha and high values of beta, the points in the likelihood space change their distribution and 'move' around the zero-one loss decision function (bisecting line third quadrant, green line). This particular behaviour can be explained by the fact that for $\alpha = \beta = 1$ we are actually giving as input a uniform distributed prior which is very unlikely in real situations; in other words, we are saying that any value for the parameter $\theta_{f|c}$ is equally probable. Instead, it is much

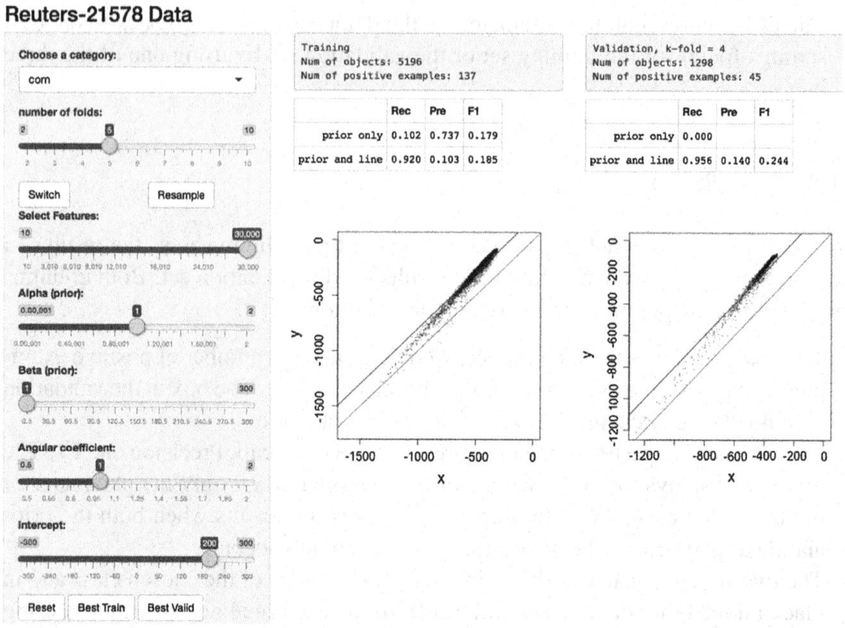

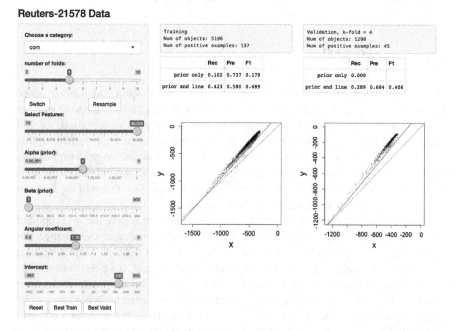

Fig. 7 Interactive text categorisation. Increase the value of intercept to recover the disproportion of the two classes

Fig. 8 Interactive text categorisation with R. Adjust angular coefficient to decrease the number of false positives

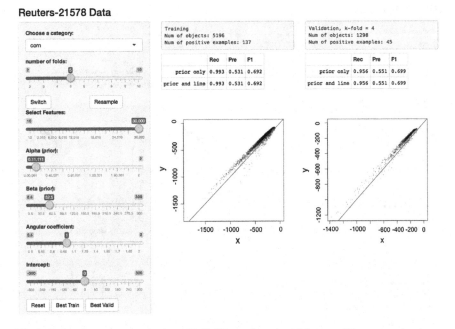

Fig. 9 Interactive text categorisation with R. Change the value of the smoothing parameters to see how points move around the zero-one loss function

more likely to observe a very small value close to zero. This is expressed by a beta function whose parameters have the values suggested in the figures.

This incremental process, as the interactive machine learning approach suggests, can significantly improve the initial results of the classifier. With this interactive application, we can also show how overfitting may generate very poor classifiers. This situation is shown in Fig. 10, where we set the alpha and beta values to their extremes and slightly adjusted the intercept and the angular coefficient to obtain an almost perfect score on the training data ($F_1 = 0.956$). With these parameters, the performance on the validation set is very low. Compared to Fig. 9, the F_1 score decreased from 0.7 to 0.4.

5   Final Remarks and Future Works

In this chapter we have presented a geometrical interpretation of likelihood spaces and an interactive text categorisation problem that makes use of this interpretation. We have explained the possible relations that exist between likelihood spaces and Bayesian Decision Theory; moreover, we have derived the same interpretation of the two-dimensional logarithmic space from the definition of classification in terms of the logit function. The interactive application shows, in a real machine learning set-

Reuters-21578 Data

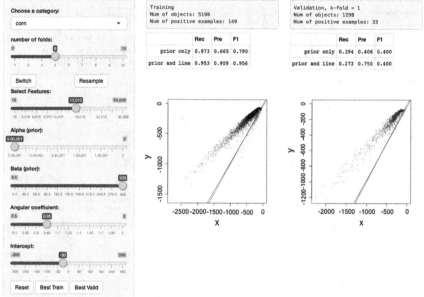

Fig. 10 Interactive text categorisation. Example of overfitting with an almost perfect score on the training data

ting, how human pattern recognition capabilities can immediately steer the learning algorithm towards one possible solution.

The importance of the visualisation approach becomes more evident when the result is used as input for the optimisation of a classifier. Theoretically, we could find the solution found with the interactive approach (if not a better one) by means of a classical full-automatic machine learning approach that searches for the best combination of parameters. The problem is that the space of the vector of parameters is huge. Although a reduction of the space can be obtained with a correct interpretation of the problem in geometrical terms [7, 10], the interactive approach can be crucial in setting the initial parameters of the function that optimises the automatic classifier.

From a theoretical point of view, there are interesting open questions about the meaning of the decision line found in the likelihood space. In particular, whether the solution has an equivalent form in the data space and in Decision Theory in general, or whether the new solution defines a completely new decision theory in the data space. Another important aspect that was not discussed in this chapter is that the smoothing parameters α and β should be optimised for each single feature instead of being equal for all the features. This problem alone would require a completely different user interface, or, in terms of classical machine learning, a study on how to choose parameters individually in an efficient way.

References

1. Amershi, S., Cakmak, M., Knox, W.B., Kulesza, T.: Power to the people: the role of humans in interactive machine learning. AI Mag. **35**(4), 105–120 (2014). http://www.aaai.org/ojs/index.php/aimagazine/article/view/2513

2. Ankerst, M., Elsen, C., Ester, M., Kriegel, H.P.: Visual classification: an interactive approach to decision tree construction. In: Proceedings of the Fifth ACM SIGKDD International Conference on Knowledge Discovery and Data Mining, KDD '99, pp. 392–396. ACM, New York, NY, USA (1999). doi:10.1145/312129.312298

3. Ankerst, M., Ester, M., Kriegel, H.P.: Towards an effective cooperation of the user and the computer for classification. In: Proceedings of the Sixth ACM SIGKDD International Conference on Knowledge Discovery and Data Mining, KDD '00, pp. 179–188. ACM, New York, NY, USA (2000). doi:10.1145/347090.347124

4. Behrisch, M., Korkmaz, F., Shao, L., Schreck, T.: Feedback-driven interactive exploration of large multidimensional data supported by visual classifier. In: 2014 IEEE Conference on Visual Analytics Science and Technology (VAST), pp. 43–52. IEEE Computer Society Press (2014). doi:10.1109/VAST.2014.7042480

5. Caruana, R., Niculescu-Mizil, A.: An empirical comparison of supervised learning algorithms. In: Proceedings of the 23rd International Conference on Machine Learning, ICML '06, pp. 161–168. ACM, New York, NY, USA (2006). doi:10.1145/1143844.1143865

6. Crestani, F., Lalmas, M., Van, Rijsbergen C.J., Campbell, I.: Is this document relevant? Probably. A survey of probabilistic models in information retrieval. ACM Comput. Surv. **30**(4), 528–552 (1998). doi:10.1145/299917.299920

7. Di Nunzio, G.: Using scatterplots to understand and improve probabilistic models for text categorization and retrieval. Int. J. Approx. Reason. **50**(7), 945–956 (2009)

8. Di Nunzio, G.: A new decision to take for cost-sensitive Naïve Bayes classifiers. Inf. Proc. Manag. **50**(5), 653–674 (2014). doi:10.1016/j.ipm.2014.04.008

9. Di Nunzio, G.: Visual classification. In: Aggarwal, C.C. (ed.) Data Classification: Algorithms and Applications, pp. 607–632. CRC Press, London (2014b)

10. Di Nunzio, G., Micarelli, A.: Pushing "underfitting" to the limit: learning in bidimensional text categorization. In: Proceedings of the 16th European Conference on Artificial Intelligence, ECAI'2004, Including Prestigious Applicants of Intelligent Systems, PAIS 2004, Valencia, Spain, pp. 465–469, 22–27 August 2004

11. Di Nunzio, G., Sordoni, A.: How well do we know Bernoulli? In: Proceedings of the 3rd Italian Information Retrieval Workshop, Bari, Italy, pp. 38–44, 26–27 January (2012). http://ceur-ws.org/Vol-835/paper5.pdf

12. Domingos, P., Pazzani, M.: On the optimality of the simple Bayesian classifier under zero-one loss. Mach. Learn. **29**(2–3), 103–130 (1997). doi:10.1023/A:1007413511361

13. Duda, R.O., Hart, P.E., Stork, D.G.: Pattern Classification, 2nd edn. Wiley, London (2000)

14. Elkan, C.: The foundations of cost-sensitive learning. In: Proceedings of the 17th International Joint Conference on Artificial Intelligence, IJCAI'01, vol. 2, pp. 973–978. Morgan Kaufmann, San Francisco, CA, USA (2001). http://dl.acm.org/citation.cfm?id=1642194.1642224

15. Fails, J.A., Olsen, D.R. Jr: Interactive machine learning. In: Proceedings of the 8th International Conference on Intelligent User Interfaces, IUI '03, pp. 39–45. ACM, New York, NY, USA (2003). doi:10.1145/604045.604056

16. Kucher, K., Kerren, A.: Text visualization browser: a visual survey of text visualization techniques. In: IEEE Information Visualization (InfoVis'14), Paris, Poster Abstract (2014)

17. Kulesza, T., Burnett, M., Wong, W.K., Stumpf, S.: Principles of explanatory debugging to personalize interactive machine learning. In: Proceedings of the 20th International Conference on Intelligent User Interfaces, IUI '15, pp. 126–137. ACM, New York, NY, USA (2015). doi:10.1145/2678025.2701399

18. Mitchell, T.M.: Machine Learning, 1st edn. McGraw-Hill, New York (1997)

19. Mladenic, D., Grobelnik, M.: Feature selection for unbalanced class distribution and Naïve Bayes. In: Proceedings of the Sixteenth International Conference on Machine Learning, ICML '99, pp. 258–267. Morgan Kaufmann, San Francisco, CA, USA (1999). http://dl.acm.org/citation.cfm?id=645528.657649

20. Neyman, J., Pearson, E.S.: On the problem of the most efficient tests of statistical hypotheses. Philos. Trans. R. Soc. Lond. Ser. A **231**, 289–337 (1993)

21. Sebastiani, F.: Machine learning in automated text categorization. ACM Comput. Surv. **34**(1), 1–47 (2002). doi:10.1145/505282.505283

22. Shneiderman, B.: Designing the User Interface: Strategies for Effective Human-Computer Interaction, 3rd edn. Addison-Wesley, Boston (1997)

23. Shneiderman, B.: Inventing discovery tools: combining information visualization with data mining. Inf. Vis. **1**(1), 5–12 (2002). doi:10.1057/palgrave/ivs/9500006

24. Singh, R., Raj, B.: Classification in likelihood spaces. Technometrics **46**(3), 318–329 (2004). doi:10.1198/004017004000000347

25. Sokolova, M., Lapalme, G.: A systematic analysis of performance measures for classification tasks. Inf. Process. Manag. **45**(4), 427–437 (2009). doi:10.1016/j.ipm.2009.03.002

26. Tan, P.N., Steinbach, M., Kumar, V.: Introduction to Data Mining, 1st edn. Addison-Wesley, Boston (2005)

27. Ware, M., Frank, E., Holmes, G., Hall, M., Witten, I.H.: Interactive machine learning: letting users build classifiers. Int. J. Hum.-Comput. Stud. **56**(3), 281–292 (2002). http://dl.acm.org/citation.cfm?id=514412.514417

28. Webb, G.I., Pazzani, M.J.: Adjusted probability Naïve Bayesian induction. In: 11th Australian Joint Conference on Artificial Intelligence Advanced Topics in Artificial Intelligence, AI '98, Brisbane, Australia, Selected Papers, pp. 285–295, 13–17 July 1998. doi:10.1007/BFb0095060

29. Yuan, Q., Cong, G., Thalmann, N.M.: Enhancing Naïve Bayes with various smoothing methods for short text classification. In: Proceedings of the 21st International Conference Companion on World Wide Web, WWW '12 Companion, pp. 645–646. ACM, New York, NY, USA (2012). doi:10.1145/2187980.2188169

Mining Movement Data to Extract Personal Points of Interest: A Feature Based Approach

Marco Pavan, Stefano Mizzaro and Ivan Scagnetto

Abstract Due to the widespread of mobile devices in recent years, records of the locations visited by users are common and growing, and the availability of such large amounts of spatio-temporal data opens new challenges to automatically discover valuable knowledge. One aspect that is being studied is the identification of important locations, i.e. places where people spend a fair amount of time during their daily activities; we address it with a novel approach. Our proposed method is organised in two phases: first, a set of candidate stay points is identified by exploiting some state-of-the-art algorithms to filter the GPS-logs; then, the candidate stay points are mapped onto a feature space having as dimensions the area underlying the stay point, its intensity (e.g. the time spent in a location) and its frequency (e.g. the number of total visits). We conjecture that the feature space allows to model aspects/measures that are more semantically related to users and better suited to reason about their similarities and differences than simpler physical measures (e.g. latitude, longitude, and timestamp). An experimental evaluation on the GeoLife public dataset confirms the effectiveness of our approach and sheds some light on the peculiar features and critical issues of location based systems.

1 Introduction

In recent years, the increasing pervasiveness of mobile devices and the ever growing mobile technologies have made location-acquisition systems available to everyone. Moreover, such systems can be easily embedded in popular apps and services, being very often active during many users' daily activities. This evolution allows to collect

M. Pavan (✉) · S. Mizzaro · I. Scagnetto
Department of Mathematics and Computer Science, University of Udine,
Via Delle Scienze 206, Udine, Italy
e-mail: marco.pavan@uniud.it

S. Mizzaro
e-mail: mizzaro@uniud.it

I. Scagnetto
e-mail: ivan.scagnetto@uniud.it

C. Lai et al. (eds.), *Information Filtering and Retrieval*,
Studies in Computational Intelligence 668, DOI 10.1007/978-3-319-46135-9_3

large datasets with spatio-temporal information, and in particular it has increased the interest of researchers on studies about user movements, behaviors and habits. Nowadays several mobile applications have been developed with the aim to exploit information extracted from raw location data. Some of those track users movement during sport activities in order to monitor their performance and to give suggestions about the next training. Other applications use GPS data to track users current position for navigation systems. Some companies use location data as a feature for social network based applications, in order to give new services to users based on their check-ins. Well known examples are Foursquare [8] that bases its entire service on users location information to give suggestions about points of interests, and Facebook [7] and Twitter [28] that allow users to add their location while posting a new message on their account, in order to add more information for other users.

The spread and popularity of this kind of mobile apps give people the possibility to track their location data in a lot of different ways, also associated to useful services, and to share with their friends this increasingly important source of information. This activity of sharing data provides in turn the additional advantage of improving the shared services offered to the community.

With these premises it is clear that there is a new important source of potentially interesting information to exploit. Whence, it is of utmost importance to design and implement an effective extraction process to get the right information from the collected raw location data. Moreover, it can be useful to envisage some post-process analysis, in order to infer additional knowledge about users. A good starting point is to recognize *important locations* for the users, i.e. *personal places of interest* (PPOIs): such places can tell a lot about their daily behavior and habits. In other words, PPOIs are places which have particular meaning for users, such as home, work, or any place where they spend a considerable amount of time during the day or which they visit frequently.

In this chapter, we focus on a novel proposal for PPOIs identification: in particular, we pay attention on how users move during their daily activities, in order to recognize the importance of places they visit according to different points of view, such as the frequency or intensity of visits. Indeed, we observed that some meaningful locations are related to users' main activities, thus they spent a lot of time in specific delimited geographic areas, such as their office or home. Other locations, instead, have been visited several time during the analyzed days, but with not the same intensity as home or office. An example of this kind of places may be the newsstand or the supermarket. In order to recognize PPOIs, we must first be able to detect the so-called *stay points* (SPs), i.e. locations where the users "may stay for a while" (see [18]). Not all stay points can be considered important places, but they are good candidates and effective off-the-shelf tools are available to extract them from raw data (whatever the source, like, e.g. a GPS-device). The candidate stay points need then to be filtered to provide the final set of PPOIs. We remark here that our proposal is technology-independent, being based only on raw data: neither we carry out any enrichments of positional data nor we use any external knowledge sources (like, e.g. georeferenced posts or resources published on Twitter, Facebook or other social networks). As we will see in Sect. 2, Urban computing [32] and trajectory data mining [37] are two research fields which can greatly benefit from this kind of work.

In the literature, earlier approaches focus on the density of detected positions inside a delimited area, and on time thresholds to check when changing area, in order to recognize the locations which might have particular meaning for users. However, this is not enough to ensure a good selection, which should also take care to discard all "false important places" (e.g. crossing at intersections or stops at traffic lights) and, at the same time, should not miss relevant locations. Indeed, grid systems which exploit density, but are based on cells of fixed dimensions, cannot always guarantee a correct recognition due to the location distribution on the geographic space: the cell bounds might overlap an important place and, as a consequence, the latter will be divided and wrongly processed as two or more distinct places.

Further complexity comes into play since users movements are affected by other factors, such as speed/acceleration, heading, relations between locations, and also by the changes of the accuracy of GPS devices during subsequent detections. Many approaches considering the speed parameter tend to identify stay points when the measurement of speed is (nearly) zero. However, this assumption is again not enough accurate (it is sufficient to think, e.g. of a walk in a park). Therefore, to properly understand users behavior and habits it seems more appropriate to analyze their movements by considering a set of combined elements to infer the right information about the way they move.

On this basis the novelty of our approach aims at overcoming the above mentioned issues and at refining the whole identification process. First of all, our method is modular; we exploit some state-of-the-art algorithms to do an initial filtering of the raw positional data. Then, we carry out a deeper analysis, taking into account some user-related measures as further steps to refine the recognition task. Namely, we consider the area covered by a stay point, the time spent in a given location and the frequency of visits. As we will see in Sect. 5, this second phase improves the final outcome in terms of precision (paying a little cost in terms of recall). In particular, our approach allows us to infer a description of places in terms of a set of *features* more related to users routine activities. Mapping the physical locations into an abstract space based on those features helps us to carry on a deeper analysis which allows us to observe if a place is repeatedly visited. Moreover, we can identify locations (e.g. rendez-vous points, newsstands, bus stops to name a few) which are visited several times during a longer period, but not with a sufficient "intensity" to be found by previous techniques.

This chapter is structured as follows: in Sect. 2 we discuss related work. The problem statement we focus on, together with the main notions and definitions, is presented in Sect. 3, while our proposed approach is described in Sect. 4. Section 5 is devoted to the experimental evaluation and, finally, we draw conclusions and some future work directions in Sect. 6.

2 Related Work

2.1 Human Mobility

Some authors focus on analyzing patterns in mobile environments. A study, presented by Laxmi et al. [16], analyzes the behavior of user patterns related to existing works from the past few years. Noulas et al. [25] analyze a large dataset from Foursquare in order to observe user check-in dynamics and find spatio-temporal patterns. Their results are useful to study user mobility and urban spaces. In this direction other authors present their work on analysis of user communities in order to build human mobility models. Karamshuk et al. [15] survey existing approaches to mobility modeling. Hui et al. [12] propose a system to improve the understanding of the structure of human mobility by analyzing the community structure as a network. Mohbey et al. [22] propose a system based on mobile access pattern generation which has the capability to generate strong patterns between four different parameters, namely, mobile user, location, time and mobile service. They focus on mobile services exploited by users and their approach shows to be very useful in the mobile service environment for predictions and recommendations. Zheng et al. [33, 35, 36] developed a brand new social network system based on user locations and trajectories, called GeoLife, which aims to mine correlations between them.

Other researchers focus on locations analysis for destination and/or prediction of places of interest (POIs); Avasthi et al. [1] propose a system for user behavior prediction based on clustering. They analyze the differentiated mobile behaviors among users and temporal periods simultaneously in order to make use of clusters and find similarities. Zheng et al. [34] perform two types of travel recommendations by mining multiple users' GPS traces: top interesting locations and locations which match user's travel preferences. In [20] the authors combine hierarchical clustering techniques, to extract physical places from GPS trajectories, with Bayesian networks (working on temporal patterns) and custom POIs databases to infer the semantic meaning of places. Thus, they are able to discover in an effective way users PPOIs. Scellato et al. [26] developed a framework called NextPlace, a novel approach to location prediction based on time of the arrival and time that users spend in relevant places. Liu et al. [19] propose a novel POI recommendation model, exploiting the transition patterns of users' preference over location categories, in order to improve the accuracy of location recommendation. Another work in the direction of providing personalized (i.e. more accurate) POI recommendations is [3] where personalized Markov chains and region localization are used to take into account the temporal dimension and to improve the performance of the system. Finally, in [9] Gao et al. leverage on content information available in location-based social networks, relating it to user behaviour (in particular to check-in actions), to improve the performance of POI recommendation systems.

2.2 Important Places Recognition

It is clear how one of the most important issues underlying these systems is the inference of users' important places. Several studies focus on this topic to propose new approaches on important places recognition, and thus provide novel algorithms to use on more complex systems. Passing from raw information about coordinates to semantically enhanced data (landmarks or places) is an important aspect in the task of discovering important places. In [14], Kang et al. introduce a time-based clustering algorithm for extracting significant places from a trace of coordinates; moreover, they evaluate it using real data from Place Lab [27]. Hightower et al. [11] exploit WiFi and GSM radio fingerprints (collected by mobile devices) to automatically discover the places people go, associating names and/or semantics to coordinates, and detecting when people return to such places. Their BeaconPrint algorithm, according to the authors, is also effective in discovering "places visited infrequently or for short durations". De Sabbata et al. [5, 6] provide an adaptation of the well-known PageRank algorithm, in order to estimate the importance of (square) locations on the basis of their geographic features (i.e. if they are contiguous or not) and the movements of users. In particular, in the calculus of the importance (rank) of a location, the speed can be used to highlight either places where the user has stopped or places where there is a high traffic density. Thus, the notion of importance of a location can be "customized" on the basis of the current needs or situation.

Li et al. [17] mine single user movements in order to identify stay points where users spend time; then, by analyzing space and time thresholds, they compute a similarity function between users based on important places that represent them. Montoliu et al. [23, 24] propose a system based on two levels of clustering to obtain places of interest: first, a time-based clustering technique which discovers stay points, then a grid-based clustering on the stay points to obtain stay regions. Isaacman et al. [13] propose new techniques based on clustering and regression for analyzing anonymized cellular network data usage to identify generally important locations.

Many of these approaches base their algorithms on the number of user detected positions within a geographic area, and in some works with attention to the elapsed time between a detected position and the next one. For instance, in [29], Umair et al. introduce an algorithm for discovering PPOIs, exploiting a notion of "stable and dense logical neighborhood" of a GPS point. The latter is automatically determined using a threshold based approach working on space, time and density of detections. To improve the recognition process, other factors and parameters are taken into consideration to enhance the algorithms. Xiao et al. [30] add semantics to users' locations based on external knowledge (POIs databases), in order to understand user's interests and compute a similarity function between two of them without overlaps in geographic spaces. More recently Bhattacharya et al. [2] extract significant places exploiting speed and the bearing change during user movement.

An interesting approach is presented in [10] where Hang et al. present Platys, an adaptive and semisupervised solution for place recognition. Its novelty amounts to the fact that it makes minimal assumptions about common parameters (e.g. types and frequencies of sensor readings, similarity metrics) which are usually tuned up manually in other systems. Instead, Platys assumes that the user visits important places sufficiently often, letting him to label the place at any time (the user is also prompted at random intervals by the system).

3 Problem Statement

3.1 Definitions

By observing a dataset of users' movements readings, it is possible to notice some coordinates where people remain stationary for long time periods, often inside buildings or delimited areas where they perform their daily activities. We call those locations *Stay Points* (SPs), as described and defined in [17]. Theoretically, a stationary user generates the same location data for all the stay time, i.e. the same point in the geographic space; we call those places *Natural Stay Points*, due to the nature of data that does not require any particular processing to understand the corresponding visited locations. However, in real situations there are several factors that affect the tracking of user movements. Due to technology limitations, there may be locations where the position detection is not possible, or the user moves in a way that the detection result cannot be so accurate. For instance, if we use a GPS, there are places where there is no signal or where the accuracy is very low due to the transportation mode that varies from underground to surface. These issues led us to have data generated by several detections which do not properly match when the user is stationary. Instead, they yielded a group of points corresponding to a location with a high density of detections within a given (limited) range. This situation may also occur when users move inside a delimited area, such as their work place where they may move among offices, or during a walk inside a mall. As described in [17], for both these latter situations we can compute the mean point of that cluster of detections in order to determine the user's stay point. We call this kind of places *Computed Stay Points*, since they approximate the original real locations. Figure 1 shows an example of user's movement readings with the two types of stay points described above. The process to identify stay points from user movements readings helps to get the set of visited locations, but neither necessarily all of them are important for the user [17] nor they provide information. By analyzing just the density of detections, some locations may be recognized as stay points even if they are not strictly related to user's main visited places. Figure 2 shows how a road crossing, where users transit a lot of time during their activities, can generate a geographic region with high density of detections, and consequently a possible stay point. We call those places *False Stay Points*, because they identify locations that do not represent a user activity, and do not provide important information about user habits and behavior.

Fig. 1 Stay point types from
user positional data

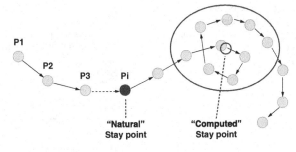

Fig. 2 The density problem
of important places
discovering

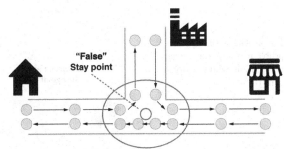

On this basis it is clear what kind of locations we consider *Important Stay Points*, namely PPOIs: locations that can help to infer information about the user who has visited them, in particular the activities that may have been carried out at each location, the stay time, and how frequently it has been visited.

3.2 Challenges and Motivations

To better understand what are the main problems and difficulties emerging with important places recognition, we list a set of conceptual problems presented by the current state-of-the-art solutions. We have also run a preliminary experiment to analyze how much the conceptual problems do appear in practical scenarios, and which of them are addressed by the existing solutions; we discuss the results in detail in Sect. 3.3.

A first approach based on density may exploit a spatial subdivision of the territory where user moved to recognize the most visited locations, and consequently assign an importance value to them, but, as described in [17], this grid-based solution is affected by several issues. The cell definition during the spatial subdivision is not a technique that can be adapted to each case and to each user movement style. As represented in Figs. 3 and 4 the cell might have a size not appropriate to analyze each user and each movement, causing the not-proper recognition of PPOIs due to what we call *boundary problem*, which might divide them (Fig. 3), or include more than one of them (Fig. 4). So two first conceptual problems are:

Fig. 3 The boundary problem of important places recognition, with too small cells (P1a)

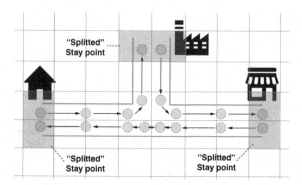

Fig. 4 The boundary problem of important places recognition, with too large cells (P1b)

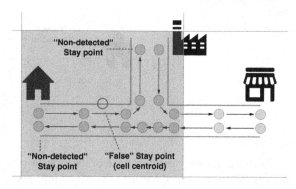

P1a: Boundary problem—undersized cells With a grid based approach, cells can be too small, and thus wrongly split a stay point.

P1b: Boundary problem—oversized cells With a grid based approach, cells can be too large, and thus wrongly identify false stay points that either merge two or more stay points, or even are created without any real stay point.

The technique used in [17, 36] for stay point computation avoids the static approach used in the grid-based solution, which mainly analyzes the user movements as an overview on a map, in favor of a dynamic approach that scans each detected position, in order to reproduce the user movements and get more information from user behavior. By using a dataset composed of users' GPS detected positions, it is possible to avoid problems related to grid cell size by focusing on defining thresholds, based on space and time, to recognize when users move and when they remain stationary.

Let $P = \{p_1, p_2, \ldots, p_n\}$ the list of points corresponding to GPS readings ordered by time of detection, tT the time threshold and dT the distance threshold. By checking the time and distance thresholds between the point p_i and the point p_{i+1} it is possible to know if the user moved or not in that specific delimited geographic region. We call that space *segment*, since it approximates the original real user movement between the two analyzed points. If the user remains stationary (i.e. she does not exceed both the thresholds), this process can be repeated by keeping fixed p_i, scanning the next points $\{p_{i+2}, p_{i+3}, \ldots, p_n\}$ and stopping when the thresholds are exceeded, in order to detect when and where the user changes behavior. At the end of this process it is possible to compute a *Mean Stay Point* based on the current set of analyzed points from p_i to p_{i+k}, with $1 \leq k \leq (n - i)$, by calculating the average latitude and longitude of points.

This technique, based on space and time thresholds, is not affected by the issues related to the cell size, which is dynamically determined, but some problems are still present (and we will indeed observe its performance in our preliminary experiment in Sect. 3.3). On straight and long trajectories, where users move with no particular changes in speed, the dynamic approach performs a scanning which, after a certain number of points, computes a stay point based on the exceeded thresholds, i.e. the mean of points in the analyzed segment, and it repeats this process for all the trajectory length, thereby determining a set of consecutive false stay points. Figure 5 shows an example of this issue displaying a path between two important places segmented with false stay points. We call this problem:

P2: segmentation problem—constant speed A trajectory between two distant important places is divided into several segments defined by the computed false stay points.

Other works in the literature [2, 33] introduce other parameters to use and improve the previous approach and minimize the segmentation problem. In particular they use new thresholds based on user speed, acceleration and even heading change, in order to better understand user behavior. More precisely, speed and acceleration

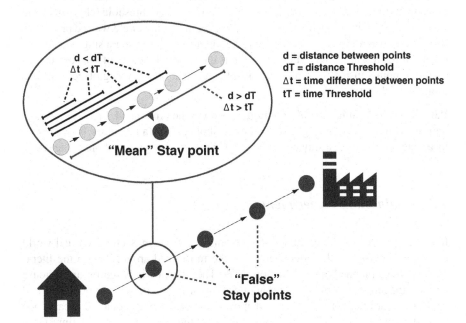

Fig. 5 The segmentation problem of important places recognition (P2)

thresholds are used in the same way as those about space and time, i.e. as soon as they are exceeded, the scanning process stops in order to compute the stay point. The heading threshold on the other hand is used in the opposite way: a constant heading indicates a movement from a SP to the next one. However, also with these approaches potentially there are some difficulties to avoid the computation of false stay points (and our preliminary experiment confirms that). For example, if a user moves with a high speed for a long time, i.e. while driving on an highway, she still exceeds the speed threshold, and after a certain amount of time the other ones, causing again the segmentation problem.

Whence, fixed thresholds may not be suitable for all user movements; indeed, some settings perfectly tuned for some users may be very wrong for others. As we will see in Sect. 5, by changing the thresholds values we observed how the recognition process varied the granularity (i.e. the number and the density of stay points) of the result, providing different set of stay points. This issue causes the computation of false stay points if the thresholds are not properly set considering the current user movements to analyze. User activities which involve several vehicles and in wider areas generate different datasets compared to users that move in small regions and mainly with one mode of transportation; whence the need of different analysis. Figures 6 and 7 show two examples where wrong thresholds raise the two last problems:

P3a: Fixed thresholds problem—slow speed In Fig. 6, it is possible to see how a region (delimited by the circle) where user moved with very slow speed, differently from the rest of the tracked movement, makes the threshold-based techniques unable to properly recognize the PPOIs, due to a too high threshold for the current tracked movement. Indeed, as soon as the speed exceeds the related threshold (changing from slow to high again), the whole slow speed region inside the circle will be processed in the same way of a walk inside a building, therefore generating a single false stay point. Moreover, the latter, whose position is the result of a mean of the coordinates of all the points inside the circle, can also be put in a totally wrong place, w.r.t. the progress of the path in the region.

P3b: Fixed thresholds problem—high speed On the other hand, Fig. 7 shows how regions with high speed (higher than the threshold set), in a trajectory between two locations, generate false stay points, again due to a not proper threshold value setup.

3.3 Preliminary Experiment

In order to understand the impact of problems described in Sect. 3.2 on real world user movements, and the effectiveness of the most used approaches in the literature, we have planned two evaluation tasks. The first one is based on an in-house dataset. Indeed, we built a mobile application (in two versions, for both iOS and Android smartphones) to gather real movement data from people. Basically, we needed a sequence of GPS points consisting in latitude, longitude, speed, timestamp and accuracy, to have a trajectory that represents how and where user moved. We

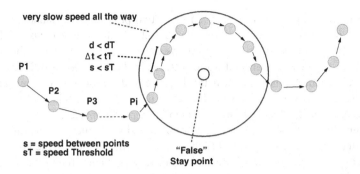

Fig. 6 The thresholds problem of important places recognition, the slow-speed issue (P3a)

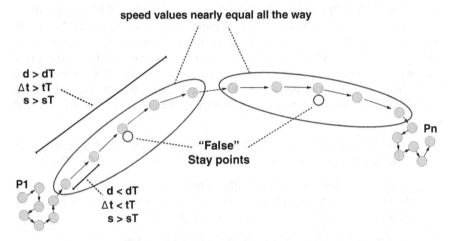

Fig. 7 The thresholds problem of important places recognition, the constant-speed issue (P3b)

have chosen a sample of 13 (Italian) users in order to collect a sufficient amount of GPS detections during 4 days of common daily activity. The second evaluation task involved the same group of 13 users, but on 4 days of movements related to 13 maps (one for each user) taken randomly from the GeoLife dataset [35]. The latter has been collected in (Microsoft Research Asia) GeoLife project by 182 users in a period of over three years (from April 2007 to August 2012: for the details see [21]).

Designing the two tasks, we paid attention to have different types of behavior, from frequent home-work travels to routines very stationary, also with different modes of transportation, e.g. motorized vehicles, bicycle, walk. We have estimated to collect (for the in-house task) and to choose (for the GeoLife task) data for 4 days for each user, in order to have enough detections to properly recognize behaviors and habits, since a lower number of days might not emphasize locations with high frequency and/or intensity.

Of course, a key difference between these two preliminary evaluation tasks is related to the users' knowledge about the datasets. In the in-house case each user evaluated the performance of the algorithms on its own data, being in the perfect condition to establish the ground truth. Instead, in the GeoLife case there was no ground truth about PPOIs available in the database, and our users were not acquainted with the Chinese regions of GeoLife. However, in each case users had the same skill and knowledge level in identifying the potential important places.

We implemented a set of popular algorithms used in the literature to check what issues affect them. The first, named G, is a static approach based on the grid method described in [17], useful to see how the boundary problem affects the results on dataset with movements from different user behaviors and habits. The second one is based on a dynamic approach and only space threshold, named S, as described in [18]. We have also implemented the T and V versions of threshold-based algorithms, since they have been often used in literature, even recently [17, 23, 29, 36]. Moreover, we have developed further versions of the latter algorithms with more parameters as thresholds, such as acceleration, and heading change, named A and H, respectively (like in [2, 33]), to see how the addition of parameters affects the PPOIs identification.

We have run all algorithms to see the results on our datasets and make some considerations about the issues explained in Sect. 3.2. Observations on results showed that:

- the static approach, namely the grid-based clustering, got variable performance due to different types of movement that need different cell sizes (P1a, P1b):
  - smaller cells allow us to recognize the right SPs, but adding a lot of false SPs;
  - larger cells generate the right number of SPs, but with wrong locations since the centroid is taken as the mean of all points in the cell;
- dynamic approaches fit well any type of movements readings;
- generally, to add new thresholds based on new parameters helped to discard false stay points;
- acceleration seems to be a too strict parameter, since too many points are discarded;
- heading change gives a low contribution to PPOIs identification, anyway it helps to improve the precision of the recognition process;
- the segmentation problem (P2) is still present;
- to use fixed thresholds does not allows us to always have a perfect setup for all situations, due to the different types of movements (P3a, P3b);
- generally, the preliminary experiment encourages us to adopt dynamic approaches exploiting several parameters with an automatic thresholds computation methodology (also helping to deal with "sensitive" parameters like, e.g. acceleration).

This preliminary experiment helped us to confirm how the above mentioned approaches still present some issues and could be improved. Figure 8 illustrates the cumulative rating distribution for all the algorithms considered in the preliminary experiment (notice that in the figure the lines for S and T algorithms coincide, the same for A and AH). Table 1 shows the average ratings, precision, recall and F-measure reported by each algorithm. We can see that, despite the higher precision

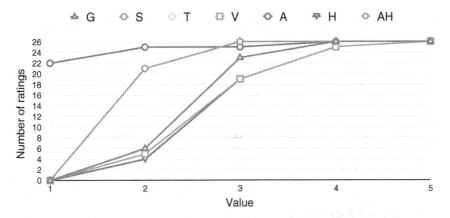

Fig. 8 Preliminary experiment: cumulative rating distribution for all algorithms

Table 1 Preliminary experiment: algorithms comparison

Algorithm	Average rating	Precision	Recall	F-measure
G	2.88	**0.186**	0.806	**0.302**
S	2.19	0.026	1	0.051
T	2.19	0.041	1	0.078
V	**3.11**	0.138	0.846	0.238
A	1.23	0.125	0.063	0.084
H	**3.15**	0.160	0.835	0.269
AH	1.23	0.125	0.063	0.084

and F-measure of *G* w.r.t. *V* and *H*, users have preferred the latter two algorithms with better average ratings (3.11 for *V* and 3.15 for *H* vs. 2.88 for *G*). This can be explained considering that *G* does not discard any candidate SPs, but it simply clusterizes them. Hence, the user can be confused looking at the representation in the map, seeing many "spurious" points scattered around in a uniform way. Moreover, sometimes the grid-based approach does not identify the right coordinates of important places, due to the cluster centroid which is affected by the high number of points contained in the cell (which can be too large). Algorithms S and T got the highest recall with a score equal to 1, but with very low performances in terms of precision and average rating. Finally, we ran the Wilcoxon test in order to verify if there are significant differences among the rating distributions got by the algorithms. The resulting p-values appear in Table 2. We can observe that there are statistical significances between several pairs of algorithms (where the *p* value <0.005). In particular, we can confirm again that increasing the number of parameters used as thresholds by the algorithms allow us to get a significant improvement, apart the cases of the threshold *T* which does not give any contribution and the threshold *H* which contributes slightly.

Table 2 Preliminar Wilcoxon test: p values

	G	S	T	V	A	H
G	–	–	–	–	–	–
S	2.467e−05	–	–	–	–	–
T	2.467e−05	NA	–	–	–	–
V	0.01966	1.507e−05	1.507e−05	–	–	–
A	4.732e−06	3.69e−06	3.69e−06	4.1e−06	–	–
H	0.01073	9.044e−06	9.044e−06	1	3.586e−06	–
AH	4.732e−06	3.69e−06	3.69e−06	4.1e−06	NA	3.586e−06

4 Proposed Approach

The key contribution and novelty of our methodology for the recognition of PPOIs
ultimately rely on a mapping from the physical space (determined by raw positional
data) to an *abstract* space, called *features space*. The latter allows us to consider
as coordinates some *features* which are semantically related to users' habits and
behaviours. Figure 9 shows an example of how important places can be positioned
in our features space. The icons indicates some kind of common places, such as
home, office, bar, mall and fuel station, but also a park, a travel to a city, or a visit
to a museum. From a pragmatic point of view places represented in this space allow
one to infer some similarities among them. In the example it is possible to see how
they may be related according to a couple of features. For instance, a bar and a fuel
station have low values on *area dimension* and *intensity* but they differ in *frequency*.
Or, home and bar may have the same *area dimension* and the same *frequency*, but
people spend a different amount of time in those locations.

Fig. 9 Some kind of
important places positioned
into the features space

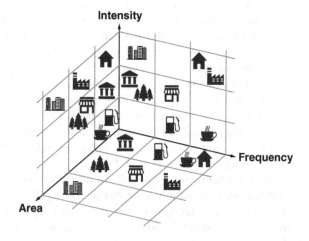

Thus, we can provide a deeper and more meaningful representation of PPOIs: for instance, in the following we will see how we can observe if a place is repeatedly visited or if it is visited several times during a longer period, but not with a sufficient intensity to be taken into consideration by previous techniques (this can be the case of, e.g. rendez-vous points, newsstands, bus stops etc.).

Our approach for PPOIs recognition consists in a method that analyzes a dataset of user movements readings, regardless of the type of technology used for tracking, even without any help from external knowledge sources. A point in the dataset just needs to be described by a set of coordinates to identify the location into a space, and a timestamp to understand the temporal order of detected points. On this basis it is possible to work on datasets with data gathered with various technologies, for instance WiFi triangulation inside a building, such as a mall or a museum, or by using a GPS for outdoor movements.

More precisely, we rely on a dynamic approach used in the literature (see [17, 23, 36]) which analyzes user movements point by point and identify stay points by checking thresholds, as described in Sect. 3. In this way we get a set of *Candidate Important Places*, due to the nature of user stay points which represent possible locations with particular meaning for the user. In more details, our approach is organized in two main steps, with a preliminary phase consisting in defining the values of the thresholds to use (based on the tracked user activity) as follows: the first one exploits a stay point computation algorithm to get a set of candidate important places; then the second step applies our feature-based technique to properly select the most important places for the current analyzed user. These steps are described in full details in the following sections.

We conclude this paragraph noticing that intensity, frequency or other similar features have already been taken into account in the literature. For instance, in [4] Chon et al. propose to combine external knowledge from crowdsourcing and social networks data, to automatically provide places with a meaningful name or a semantic meaning. In order to carry out this goal, they consider several factors such as *residence time* (indicating "stay behavior of users at a place tied with time-of-day") and *stay duration* (indicating "pattern of stay behavior without time-of-day"). More in general, the very concepts of features space and feature vector have been exploited in [31], in order to find similarities between users, starting from their location histories.

4.1 Preliminary Phase: Thresholds Definition

As preliminary phase, we address the threshold definition problem. As explained in Sect. 3, there are no fixed values for thresholds that fit perfectly for each user and for each dataset; therefore a brief reasoning may help to understand what kind of movements we are analyzing. During our preliminary experiment (see Sect. 3.3) we observed that changing the speed threshold highly affected the results for users with different use of the vehicles and transportation mode, and even the acceleration and heading change are strictly related to how users move routinely. Therefore, we define a method for extracting a good set of values for these three thresholds. We run a scan on the dataset in order to get information about the three parameters described above, paying attention on the median of non-zero values of speed, acceleration and heading change between each couple of points. This choice stems from the considerations discussed in Sect. 3; indeed we observe that the median of all speeds reached by the analyzed user, may be a good value to identify when user changes behavior. We adopt the same consideration for acceleration and heading change, in order to have a set of thresholds to use in the next step to build algorithm variants for comparison purposes.

About distance and time we keep the thresholds fixed. We set the distance threshold dT equal to 50 m, and time threshold tT equal to 50 s. These are parameters set empirically, by observing a sample of user movements during the preliminary experiment (see Sect. 3.3), where we noticed that they do not strongly affect the stay points identification.

4.2 Step 1: Stay Points Computation

As second step we identify the user stay points, by using a dynamic approach which consists in a scan of all points in the dataset, in order to simulate and reproduce the movement, and exploits thresholds based on some parameters to understand user behavior and recognize when and where users move or remain stationary in a location. To make possible a proper evaluation of our method, we implement several solutions of this dynamic approach; in particular, we want to compare earlier methods based just on space and/or time to others that also exploit speed, acceleration and/or heading change. By observing the results of our preliminary experiment (see Sect. 3.3), we notice that space and time are not sufficient to properly determine the right set of stay points, and also other related works take into account other parameters [2, 33]. Moreover, acceleration and heading change were too strict as parameters of selections, in our heterogeneous dataset, and they have led the algorithm to discard too many stay points. Based on these observations, we chose to use space, time and speed parameters as thresholds for the stay point computation module. More formally, during the analysis of a point p_i and a point p_j, i.e. the next one in the user trajectory, we add the point p_i to the list of candidates for the stay point computation if one or more of the following constraints are satisfied:

$$distance(p_i, p_j) \leq dT$$
$$timeDiff(p_i, p_j) \leq tT$$
$$speed(p_i, p_j) \leq sT$$

During the computation we may also take into account the accuracy of coordinates detected during the movement tracking process. If the analyzed dataset provides the accuracy values for each point reading, it is possible to improve the parameters computation between two points. For instance, if we use a dataset with data gathered by using a GPS, we can discard coordinates with very low accuracy, in order to avoid weird values due to detection errors, or even we can exploit the instant speed detection, if the accuracy is good enough to make the value reliable. On this basis, our method checks the presence of the accuracy parameter into each entry of the dataset in order to exploit it for discarding data with low reliability, and to use the instant speed, if detected. If the dataset provides this additional information, we keep only data with $accuracy \leq 30$ m,[1] in order to avoid errors in distance computation and user speed analysis, due to problems with point data acquisition. For the speed computation we also take into account the instant speed as follows:

$$speed(p_i, p_j) = \begin{cases} \frac{segSpeed(p_i,p_j)+iSpeed(p_i,p_j)}{2} & p_j.acc \leq 10 \\ segSpeed(p_i, p_j) & \text{otherwise,} \end{cases}$$

where $p_j.acc$[2] is the GPS accuracy value for that specific detection, $iSpeed(p_i, p_j)$ is the average value of instantaneous speed detected by the GPS in points p_i and p_j, and $segSpeed(p_i, p_j)$ is the average speed from the point p_i to the point p_j in the user trajectory, i.e. the space segment $\overline{p_i p_j}$.

If $speed(p_i, p_j)$ is above the speed threshold, the user might be moving, thus we update the point scanning with $i = j$, in order to discard locations which could not be appropriate stay points. Otherwise, if user has low speed, we keep fixed p_i and perform a scan over the next points p_{i+k}, with $1 \leq k \leq (n - 1)$, in order to detect locations to add to the list of candidates for the stay point recognition, focusing on distance and time thresholds, but also keeping checked the speed for the scan update. When the speed threshold is exceeded again, the list of candidates is processed in order to compute a *Mean Stay Point*, and the scan can continue with the next points. Algorithm 1 illustrates the stay points computation process in detail.

---

[1] $Accuracy \leq 30$ is a parameter set empirically, by observing the raw data.

[2] $p_j.acc \leq 10$ is a parameter set empirically, by observing a set of GPS detections in several signal acquisition conditions.

Algorithm 1 SPs computation

Input: A set of user movement readings $P = \{p_0, p_1, \ldots, p_n\}$, a distance threshold dT, a time threshold tT, and a speed threshold sT
Output: A set of SPs SP
1: $i, j = 0; n = |P|; q = newPoint$
2: $CP = \{p_0\}$ ▷ list of candidate points
3: $SP = \{\}$ ▷ final list of SPs
4: **while** $i < n$ **do**
5:     $j = i + 1$
6:     **while** $j < n$ **do**   ▷ $p_i, p_j \in P$
7:        **if** $dist(p_i, p_j) > dT$ & $time(p_i, p_j) > tT$ & $speed(p_i, p_j) > sT$ **then**
8:           $q.coord = meanCoordInCP()$ ▷ $\forall p_k | i \leq k < j$
9:           $q.arrivalTime = p_i.time$
10:         $q.leaveTime = p_j.time$
11:         $SP.insert(q)$
12:         $i = j$
13:         $CP = \{p_j\}$
14:         **break**
15:        **else**
16:           **if** $speed(p_i, p_j) \leq sT$ **then**
17:             $CP.insert(p_j)$
18:           **end if**
19:           $j = j + 1$
20:        **end if**
21:     **end while**
22: **end while**
23: **return** SP

Table 3 Stay points computation algorithms variants

Algo name	Thresholds
S	Space
T	Space, time
V	Space, time, speed
A	Space, time, speed, acceleration
H	Space, time, speed, heading change
AH	Space, time, speed, acceleration, heading change

With the same methods described in the algorithm for the thresholds definition and in Algorithm 1 we implemented several versions of the stay point computation algorithm based on different thresholds, in order to compare the performance and understand what are the most useful set of thresholds for user behavior analysis. Table 3 shows all variants implemented for the comparison and evaluation process.

4.3 Step 2: Important Places Recognition

The main idea inspiring this step of our method is to map physical locations to an *abstract* space defined by a set of features more semantically related to users' habits and behaviours. For instance, a candidate feature is the frequency of visits, since

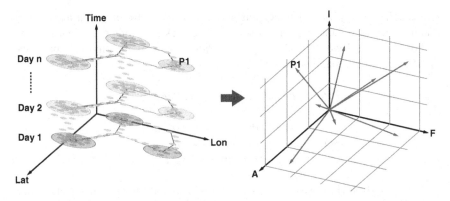

Fig. 10 The feature-based approach that moves places into the feature space

users tend to behave similarly in everyday's life. Thus, in order to define a procedure for the important places recognition, it is useful to observe users' movements across a period of time longer than a single day.[3] In other words we want to explore the possibility of superimposing the locations visited by user several times in order to extract additional semantic information, and possibly refining the results of the previous phase. Hence, to implement such strategy, we consider new parameters to describe locations, alongside latitude, longitude and timestamp, that may help to improve the recognition process.

First, we modeled the user movements readings into a three-dimensional space where a point is described by the three original raw data gathered by sensors (latitude, longitude and timestamp), in order to have a distribution of points into this space that reproduces the original user movements (see Fig. 10 (left)). We observed how the data is divided into groups, nearly in layers, which approximately represent the days when user performed the activity. Therefore this aspect makes possible further analysis and helps to get more information form each locations. On this basis we define a set of three features to describe each important place (PPOI) as a vector $PPOI = \langle A, I, F \rangle$, where A, the *Area* of the PPOI, is a value which indicates the diagonal extension of the rectangular region that spans over all points involved in the stay point computation. As explained in Sect. 3, when users visit locations tend to not stay perfectly stationary, but to move around a delimited area. We also keep the set of physical coordinates which describe it, in order to also represent it graphically for user-testing purposes, and for checking potential overlaps. The feature I, the PPOI *Intensity*, is a value which indicates how many times the user position has been detected inside the PPOI's area. Finally, the feature F, the PPOI *Frequency*,

---

[3]Otherwise, activities of a single day may escape from the usual routine and could easily hinder the recognition process.

indicates how many times that location has been visited by the user, thus a parameter that increments its value each time the user came back for another visit in that place.

Formally, we have the following map ($PhysSpace_{SP}$ is the physical space and $FeatSpace_{SP}$ is the features space):

$$feature : PhysSpace_{SP} \longrightarrow FeatSpace_{SP}$$

$$feature : \begin{array}{l} \langle sp.latitude, \\ sp.longitude, \\ sp.timestamp \rangle \end{array} \longmapsto \begin{array}{l} \langle sp.Area, \\ sp.Intensity, \\ sp.Frequency \rangle \end{array}$$

Figure 10 shows how we model movements data into the three-dimensional space, defined by *latitude*, *longitude* and *timestamp*, and how we map stay points from the physical space to the features space based on the new three dimensions A, I and F.

This feature-based approach makes possible a refinement step to emphasize locations visited intensively and/or repeatedly, and also to filter out the false stay points that the previous phase was not able to discard. In this phase we first analyze the SPs computed during the previous step, in order to calculate the stay point area A based on the detections involved in the stay point computation. Then we extract the number of detections inside that area to compute the value of I, and get the PPOI intensity, and we also set the frequency of each SP to 1 as initial value. With all SPs described into our features space we can analyze both the A and I distribution over the entire dataset, in order to better understand user behavior and define new thresholds that may help to filter out some places not important for that user. We set an area threshold aT equal to 3 km to run a pre-filtering process which discards SPs with diagonal area ≥ 3 km.[4] This operation helps to identify the kind of false stay points described in Sect. 3 as problems P3a and P3b, where a wrong threshold set may cause SPs generated by detections that span over a wide space. We also observed how a different use of vehicles and mode of transportation generate different density of detections, with the consequence of having higher I values for users that usually move slower. This issue led us to define an intensity threshold iT in order to discard SPs with I too low in proportion to the values obtained in the rest of the dataset. Moreover, with particular attention on the I values of adjacent SPs to recognize where the segmentation problem may have occurred (see Sect. 3). On this basis, we define the intensity threshold $iT = max3consec(intensities)$, where *intensities* is the array with all intensity values of each SP, and the method *max3consec* returns the maximum value of intensity that in the SPs sequence is present at least three times in a row (up to some tolerance threshold for dealing with measurement errors and small deviations[5] from the maximum value). This technique helps us to recognize

---

[4]$aT \geq 3$ is a parameter set empirically, by observing user movements during our preliminary experiment described in Sect. 3.3.

[5]For instance, the user slightly changes speed while driving along a highway.

Fig. 11 Example of overlapping activities

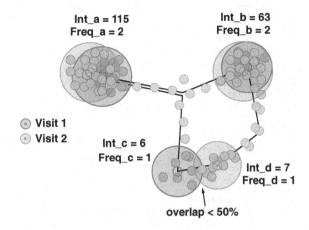

where the user is moving, and also where he is generating the same intensity values. By selecting the maximum value, we can discard the false stay points induced by the scenario described in the conceptual problem P2. Moreover, automatically computing the intensity threshold as previously described, we avoid locations where users stopped just once and for an amount of time not so remarkable as the time spent in home, office, supermarket, etc. Such places may be intersections with traffic lights which block vehicles for a long time, traffic-clogged streets, or rail crossings. Based on these two thresholds we run a pre-filtering process, to have a more accurate subset of SPs and proceed to take into account the frequency of visits.

By analyzing SPs sequentially, by timestamp, it is possible to check if their rectangular areas overlap, in order to get information about locations visited repeatedly. If the areas of two locations overlap with an intersection region $\geq 50\,\%$[6] of one of the current analyzed areas, they may be considered to represent the same place. In Fig. 11 it is possible to see an example of two visits on the same geographic area where for the locations a and b there are very similar detections on both days, therefore they represent the same important place. In that case the intensity values will be summed, the area will be their union, and the frequency will get a value equal to 2 because of the number of visits. Otherwise, the locations c and d have detections with an area overlap $<50\,\%$, therefore they will be considered as two separated places. After this filtering step, we repeat the process of merging areas several times until we get just separated regions, which identify our important places. As final phase, we run again the filtering process in order to clean out PPOIs that may have been generated with too large areas. All phases of our method named AIF are illustrated in Algorithm 2.

---

[6]$Overlap \geq 50\,\%$ is a parameter set empirically, by observing user movements during the preliminary experiment.

Algorithm 2 AIF computation

Input: A set of user stay points $SP = \{sp_1, sp_2, \ldots, sp_n\}$
Output: A set of important places $PPOI$
1: $aT, iT, fT = 0; q = newPoint$
2: $areas, intensities = \{\}$ ▷ arrays with all values
3: $PPOI = \{\}$ ▷ final list of important places
4: **for** sp_i in SP **do**   ▷ pre-filtering
5:      $insertInAreas(sp_i.computeArea())$
6:      $insertInIntensities(sp_i.computeIntensity())$
7:      $sp_i.freq = 1$
8: **end for**
9: $aT = 3$ ▷ empirically set to remove SPs with area diagonal > 3 km
10: $iT = max3consec(intensities)$ ▷ the maximum value repeated at least three times in a row
11: $PPOI = preFiltering(SP, aT, iT, areas, intensities)$
12: $overlaps = true$ ▷ to check overlaps during the points scan
13: **while** $overlaps == true$ **do**   ▷ points merging
14:     $overlaps = false$
15:     **for** p_i in $PPOI$ **do**
16:         **for** p_j in $PPOI \setminus \{p_i\}$ **do**
17:             **if** $overlap(p_i, p_j)$ **then**
18:                 $overlaps = true$
19:                 $q.area = mergePointsAreas(p_i.area, p_j.area)$ ▷ A
20:                 $q.intensity = p_i.intensity + p_j.intensity$ ▷ I
21:                 $q.frequency = p_i.frequency + p_j.frequency$ ▷ F
22:                 $PPOI.add(q)$
23:                 $PPOI.remove(p_i, p_j)$
24:                 break
25:             **end if**
26:         **end for**
27:         **if** $overlaps == true$ **then**
28:             break
29:         **end if**
30:     **end for**
31: **end while**
32: **for** p_i in $PPOI$ **do**   ▷ post-filtering
33:     $PPOI = postFiltering(PPOI, aT, iT, areas, intensities)$
34: **end for**
35: **return** $PPOI$

---

5 Experimental Evaluation

5.1 Experimental Design

To evaluate several approaches to important places identification, and to benchmark our proposed solution, we run a set of algorithms over the GeoLife dataset. We implemented the set of threshold-based algorithms described in Sect. 4, to make possible a comparison among the approaches used in the literature. As final algorithm, we have implemented our solution *AIF*, as defined in Sect. 4, to compare it to the other approaches. We have selected a sample of 16 people to evaluate the results, distributed as follows:

- 62 % men, 38 % women (all Italians);
- 62 % with age between 21 and 30, 38 % more than 30;
- 88 % with very good familiarity with smartphones;

- 56 % with intensive use of map services, 25 % with intermediate use and 19 % occasional user.

As in the preliminary experiment, all of them were not familiar with the geographic regions in GeoLife, due to the different nationality: GeoLife data have been collected in China, while our participants were Italians. Since the GeoLife dataset does not contain ground truth about PPOIs, this fact yielded the positive effect that all the participants had the same skill and knowledge level in identifying the potential important places.

We have defined a test protocol providing detailed instructions to participants so as to guide them during the evaluation in the definition of the aspects to take into consideration. We have implemented a testing tool for them to show on a map some randomly selected sets of GPS detections form GeoLife dataset, with attention on choosing al least four consecutive days of movements readings. The participants had available a heatmap to better understand the original user movement and properly evaluate the PPOIs showed as pins on the map. The tool displays sequentially and randomly maps with pins computed by one of the algorithms previously described, in order to make not clear to participants how to associate the algorithms with the corresponding suggestions. This is a precaution to not affect them with clues during the test. During an evaluation a number between 1 and 5 indicates how they judge the overall PPOIs identification. The meanings of the rate values are the following:

1. SPs retrieved $\leq 20\%$;
2. SPs retrieved $>20\%$ and $\leq 50\%$ or a very high number of false SPs;
3. SPs retrieved $>50\%$ and $\leq 80\%$ or $>80\%$, but with an high number of false SPs;
4. SPs retrieved $>80\%$ and $\leq 90\%$ and zero or a very low number of false SPs;
5. SPs retrieved $>90\%$ and zero or a very low number of false SPs.

Moreover, they were requested to indicate which pins properly represent PPOIs visited during the tracked activity, and also how many have been missed: this allows to compute Precision, Recall, and F-measure (i.e. the harmonic mean of Precision and Recall) of each algorithm.

5.2   Results

Results are reported in Fig. 12 and Table 4. The figure shows the cumulative distribution of the ratings obtained by each approach; the table shows, besides the average rating for each algorithm, also its precision, recall, and F-measure. The rating distribution and the average ratings show how the S algorithm obtained many 1-value ratings, due to the low filtering that it applies with the single threshold approach, thus getting a mean rate equal to 1.16; the T solution has been evaluated slightly better but most of rates still remains low; the adding of speed improved the performance as we expected; the A algorithm instead has worsened the identification process due to the acceleration parameter which has made too strict the PPOIs recognition process;

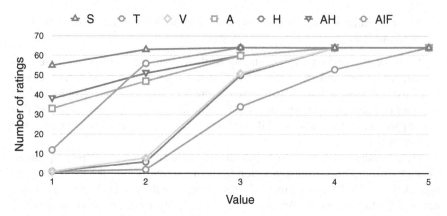

Fig. 12 Cumulative rating distribution for all algorithms for important places identification

Table 4 Algorithms comparison

Algorithm	Average rating	Precision	Recall	F-measure
S	1.16	0.004	**0.992**	0.007
T	1.94	0.009	**0.992**	0.019
V	3.06	0.126	0.657	0.211
A	1.81	0.286	0.217	0.247
H	3.11	0.131	0.657	0.219
AH	1.67	0.172	0.161	0.166
AIF	**3.59**	**0.370**	0.606	**0.459**

H, based on the heading change parameter, obtained a good performance but also a minimal improvement over V; the algorithm AH has been penalized by the use of acceleration; finally, our proposed method AIF collected a lot of positive evaluations, obtaining a mean rate equal to 3.59, the highest score among all the compared algorithms.

The simpler methods, such as S and T, got the higher recall values but with Precision very low, due to the filtering process that discards few false stay points, and provides a final set of PPOIs not so much different from the original set of movements readings. By adding more parameters as thresholds, the identification process improved, providing more accurate set of PPOIs. Moreover, the introduction of an automatic threshold algorithm computation has further improved the results. V and H solution increased the Precision, also keeping good Recall values. But the use of acceleration has reduced a lot the Precision of algorithms A and AH, obtaining very low performance in every aspect. The AIF solution has proven to be the most accurate method, with the highest precision and a good Recall, obtaining a good overall evaluation with the highest F-measure. The results confirm how AIF has improved the PPOIs identification process by providing few false stay points,

Table 5 Wilcoxon test: p values

	S	T	V	A	H	AH
S	–	–	–	–	–	–
T	1.6e−12	–	–	–	–	–
V	1.415e−13	6.343e−14	–	–	–	–
A	4.623e−07	0.1731	2.778e−11	–	–	–
H	1.109e−13	1.364e−13	0.1489	1.903e−11	–	–
AH	2.928e−06	0.004182	5.605e−12	0.003353	5.251e−12	–
AIF	1.317e−12	1.086e−12	5.844e−09	1.364e−13	2.752e−08	1.107e−13

and guaranteeing not to lose too many important locations. Moreover, it provides a less confusing visualization on the map, and it is less affected by the type of users' movements.

5.3 Statistical Significance

Moreover, we have run a statistical test to determine whether there are any significant differences between the means of ratings got by the algorithms. Due to the nature of the data with non-normal distribution we run the Wilcoxon test in order to verify if datasets have significant differences.

Table 5 shows all the resulting p-values for each couple of algorithms to compare. We can observe that the most of them got very low p-value, lower than the standard Wilcoxon threshold 0.05. Therefore, this output indicates a statistically significant difference between means, and consequently a relevant improvement in performance for those algorithms. It is possible to notice how the low performances of T and A are not statistically different. Finally, the use of the heading change threshold did not bring a significant improvement when used in algorithm V, and it provided only a small noticeable improvement when used in algorithm A.

6 Conclusions and Future Work

In this chapter we have presented our proposal of important locations identification. We have stated the most common issues related to the recognition process, then we have described our approach, that consists in a new model based on a space transition from physical space to a features space, where locations are described by a set of features more related to users' habits and behaviors. The experiment performed in this work has demonstrated that the proposed approach results more effective than other related works in terms of performance, and also in difficult situations, where other algorithms are affected by the problems described in the chapter. Moreover, the feature-based approach allows us to add more semantic value to important places, providing new information that future works may exploit for locations classification and similarity computation. For future work, we plan to work on the features space in order to explore the possibility to expand the features set and design a locations classifier based on this approach. Moreover, we want to analyze user movement types, in particular what kind of vehicles people use, and what pace they have while walking or running, in order to provide new data and further improve the important locations identification. Finally, it would be interesting to take into account the analysis of places co-located inside a single building or within a small area.

References

1. Avasthi, S., Dwivedi, A.: Prediction of mobile user behavior using clustering. In: Proceedings of TSPC13, vol. 12, p. 14th (2013)
2. Bhattacharya, T., Kulik, L., Bailey, J.: Extracting significant places from mobile user GPS trajectories: a bearing change based approach. In: Proceedings of ACM SIGSPATIAL GIS 2012, pp. 398–401. ACM (2012)
3. Cheng, C., Yang, H., Lyu, M.R., King, I.: Where you like to go next: successive point-of-interest recommendation. In: Proceedings of IJCAI'13, pp. 2605–2611. AAAI Press (2013)
4. Chon, Y., Kim, Y., Cha, H.: Autonomous place naming system using opportunistic crowdsensing and knowledge from crowdsourcing. In: Proceedings of ACM/IEEE International Conference on Information Processing in Sensor Networks (IPSN), pp. 19–30. IEEE (2013)
5. De Sabbata, S., Mizzaro, S., Vassena, L.: Spacerank: using pagerank to estimate location importance. In: Proceedigns of MSoDa08, pp. 1–5 (2008)
6. De Sabbata, S., Mizzaro, S., Vassena, L.: Where do you roll today? Trajectory prediction by spacerank and physics models. Location Based Services and TeleCartography II. LNGC, pp. 63–78. Springer, Berlin (2009)
7. Facebook check-in: Who, what, when, and now...where. https://www.facebook.com/notes/facebook/who-what-when-and-nowwhere/418175202130 (2014)
8. Foursquare check-in: About. https://foursquare.com/about (2014)
9. Gao, H., Tang, J., Hu, X., Liu, H.: Content-aware point of interest recommendation on location-based social networks. In: Proceedings of the Twenty-Ninth AAAI Conference on Artificial Intelligence (AAAI'15), pp. 1721–1727. AAAI Press (2015)
10. Hang, C.W., Murukannaiah, P.K., Singh, M.P.: Platys: user-centric place recognition. In: AAAI Workshop on Activity Context-Aware Systems (2013)
11. Hightower, J., Consolvo, S., LaMarca, A., Smith, I., Hughes, J.: Learning and recognizing the places we go. In: Proceedings of UbiComp 2005: Ubiquitous Computing, pp. 159–176. Springer (2005)
12. Hui, P., Crowcroft, J.: Human mobility models and opportunistic communications system design. Philos. Trans. R. Soc. A Math. Phys. Eng. Sci. **366**(1872), 2005–2016 (2008)
13. Isaacman, S., Becker, R., Cáceres, R., Kobourov, S., Martonosi, M., Rowland, J., Varshavsky, A.: Identifying important places in peoples lives from cellular network data. In: Pervasive Computing, pp. 133–151. Springer (2011)
14. Kang, J.H., Welbourne, W., Stewart, B., Borriello, G.: Extracting places from traces of locations. In: Proceedings of the 2nd ACM International Workshop on Wireless Mobile Applications and Services on WLAN Hotspots, pp. 110–118. ACM (2004)
15. Karamshuk, D., Boldrini, C., Conti, M., Passarella, A.: Human mobility models for opportunistic networks. IEEE Commun. Mag. **49**(12), 157–165 (2011)

16. Laxmi, T.D., Akila, R.B., Ravichandran, K., Santhi, B.: Study of user behavior pattern in mobile environment. Res. J. Appl. Sci. Eng. Technol. **4**(23), 5021–5026 (2012)
17. Li, Q., Zheng, Y., Xie, X., Chen, Y., Liu, W., Ma, W.Y.: Mining user similarity based on location history. In: Proceedings of ACM SIGSPATIAL GIS 2008, p. 34. ACM (2008)
18. Lin, M., Hsu, W.J.: Mining gps data for mobility patterns: a survey. Pervas. Mobile Comput. **12**, 1–16 (2014)
19. Liu, X., Liu, Y., Aberer, K., Miao, C.: Personalized point-of-interest recommendation by mining users' preference transition. In: Proceedings of ACM CIKM 2013, pp. 733–738. ACM (2013)
20. Lv, M., Chen, L., Chen, G.: Discovering personally semantic places from gps trajectories. In: Proceedings of ACM CIKM 2012, New York, USA, pp. 1552–1556. ACM (2012)
21. Microsoft Research Asia: GeoLife project. http://research.microsoft.com/en-us/downloads/b16d359d-d164-469e-9fd4-daa38f2b2e13/ (2012)
22. Mohbey, K.K., Thakur, G.: User movement behavior analysis in mobile service environment. Br. J. Math. Comput. Sci. **3**(4), 822–834 (2013)
23. Montoliu, R., Blom, J., Gatica-Perez, D.: Discovering places of interest in everyday life from smartphone data. Multimed. Tools Appl. **62**(1), 179–207 (2013)
24. Montoliu, R., Gatica-Perez, D.: Discovering human places of interest from multimodal mobile phone data. In: Proceedings MUM 2010, p. 12. ACM
25. Noulas, A., Scellato, S., Mascolo, C., Pontil, M.: An empirical study of geographic user activity patterns in foursquare. ICWSM **11**, 70–573 (2011)
26. Scellato, S., Musolesi, M., Mascolo, C., Latora, V., Campbell, A.T.: Nextplace: a spatio-temporal prediction framework for pervasive systems. In: Pervasive Computing, pp. 152–169. Springer (2011)
27. Schilit, B., LaMarca, A., Borriello, G., Griswold, W., McDonald, D., Lazowska, E., Balachandran, A., Hong J., Iverson, V.: Challenge: ubiquitous location-aware computing and the place lab initiative. In: Proceedings of the 1st ACM International Workshop on Wireless Mobile Applications and Services on WLAN, (WMASH 2003), San Diego, CA, September 2003
28. Twitter check-in: How to tweet with your location. https://support.twitter.com/entries/122236-how-to-tweet-with-your-location (2014)
29. Umair, M., Kim, W.S., Choi, B.C., Jung, S.Y.: Discovering personal places from location traces. In: Proceedings of ICACT'14, pp. 709–713. IEEE (2014)
30. Xiao, X., Zheng, Y., Luo, Q., Xie, X.: Finding similar users using category-based location history. In: Proceedings of ACM SIGSPATIAL GIS 2010, pp. 442–445. ACM (2010)
31. Xiao, X., Zheng, Y., Luo, Q., Xie, X.: Inferring social ties between users with human location history. J. Ambient Intell. Humaniz. Comput. **5**(1), 3–19 (2014)
32. Zheng, Y., Capra, L., Wolfson, O., Yang, H.: Urban computing: concepts, methodologies, and applications. ACM Trans. Intell. Syst. Technol. **5**(3), 38 (2014)
33. Zheng, Y., Li, Q., Chen, Y., Xie, X., Ma, W.Y.: Understanding mobility based on gps data. In: Proceedings of UbiComp'08, pp. 312–321. ACM (2008)
34. Zheng, Y., Xie, X.: Learning travel recommendations from user-generated gps traces. ACM TIST **2**(1), 2 (2011)
35. Zheng, Y., Xie, X., Ma, W.Y.: GeoLife: a collaborative social networking service among user, location and trajectory. IEEE Data Eng. Bull. **33**(2), 32–39 (2010)
36. Zheng, Y., Zhang, L., Xie, X., Ma, W.Y.: Mining interesting locations and travel sequences from GPS trajectories. In: Proceedings of the WWW'09, pp. 791–800. ACM (2009)
37. Zheng, Y., Zhou, X.: Computing with Spatial Trajectories. Springer, Berlin (2011)

SABRE: A Sentiment Aspect-Based Retrieval Engine

**Annalina Caputo, Pierpaolo Basile, Marco de Gemmis,
Pasquale Lops, Giovanni Semeraro and Gaetano Rossiello**

Abstract The retrieval of pertaining information during the decision-making process requires more than the traditional concept of *relevance* to be fulfilled. This task asks for *opinionated* sources of information able to influence the user's point of view about an entity or target. We propose SABRE, a Sentiment Aspect-Based Retrieval Engine, able to tackle this process through the retrieval of opinions about an entity at two different levels of granularity that we called aspect and sub-aspect. Such fine-grained opinion retrieval enables both an aspect-based sentiment classification of text fragments, and an aspect-based filtering during the navigational exploration of the retrieved documents. A preliminary evaluation on a manually created dataset shows the ability of the proposed method at better identify $\langle$*aspect, sub-aspect*$\rangle$ with respect to a term frequency baseline.

1 Introduction

Looking for others' opinions, impressions, and experiences is one of the first steps we usually perform when obliged to face a decision process. This could be the next president election, the booking of a room for the next holidays, or just the purchase

A. Caputo (✉) · P. Basile · M. de Gemmis · P. Lops · G. Semeraro · G. Rossiello
Department of Computer Science, University of Bari Aldo Moro, Bari, Italy
e-mail: Annalina.Caputo@uniba.it

P. Basile
e-mail: Pierpaolo.Basile@uniba.it

M. de Gemmis
e-mail: Marco.deGemmis@uniba.it

P. Lops
e-mail: Pasquale.Lops@uniba.it

G. Semeraro
e-mail: Giovanni.Semeraro@uniba.it

G. Rossiello
e-mail: Gaetano.Rossiello@uniba.it

© Springer International Publishing AG 2017   63
C. Lai et al. (eds.), *Information Filtering and Retrieval*,
Studies in Computational Intelligence 668, DOI 10.1007/978-3-319-46135-9_4

of a new product. Whatever the task, we start the process of making up our own opinion about a topic exploring both the available information and comments from others' experience. In this context, the concept of "relevance" is more than something pertaining an information need, like in a standard retrieval task. Indeed, valuable and relevant information should also bear a *subjective* point of view on a given topic or entity. Opinion retrieval (OR) aids such a process, since beyond the topical relevance of information retrieval (IR) systems, it requires documents to be opinionated.

Aspects play an important role in sentiment analysis and opinion mining. While the general sentiment or expression of opinion toward an entity is important to grasp the "overview" on a given subject, and can help during the initial investigation on a topic of interest, deeper in the process of decision-making, users are somehow more interested in specific aspects (or features) of interest. Classical examples are product reviews, where usually the user has a specific "aspect of interest" that leads her/him towards the thumb up/thumb down final decision. For example, searching for a hotel, someone may be more interested in the location, while others give more prominence to the value for money. These are perceived as different *aspect*s of the same *entity* (i.e. the target hotel). However, the extraction and organization of aspects from opinionated sources does not always match the user's interests and preferences. Usually, the assignment of aspects does not reflect the text content, but rather follows a manually created list of points of interest for a given domain. Figure 1 shows different lists of aspects from four well-known on-line booking services. The lists differ from one another, although there is some overlap, and this suggests that there is not a unique way of organizing aspects of interest for a given domain.

Generally, all entries cover broad aspects, but there is no way of further refining such a list. For example, Fig. 1a, c, d all report about "comfort". In two cases (booking.com and hotel.com), this aspect can refer to either room or hotel comfort. When referred to the room, the comfort aspect might be further refined as: bed/pillow comfort, spaciousness of the room, existence of special facilities (like Jacuzzi) or new furnitures, and so on. Moreover, since grades on aspects are given independently from the review, those aspects may not appear in the opinionated text.

This chapter describes SABRE, a Sentiment Aspect-Based Retrieval Engine, which takes into account aspects during both the process of sentiment classification of a given text and the navigation of retrieved documents (Sect. 2). Our main contribution is the aspect extraction algorithm, which processes text in a completely unsupervised manner to detect opinion bearing sentences and their subject. Aspects are organized in a two-level hierarchy in order to enable different levels of search granularity on the opinions of interest. The aspect extraction and weighing process is described in Sect. 3. Then in Sect. 4 we set forth an aspect-based retrieval model that takes advantages from the extracted aspects and their sentiment. Finally, we assess the proposed algorithm on a manually created dataset in order to validate its ability to recognize opinion-related aspects in a text at different levels of granularity (Sect. 5).

(a) booking.com (b) Venere.com

(c) Expedia.com (d) hotel.com

Fig. 1 List of aspects from different hotel booking web sites

2 SABRE

An opinion is defined as a sentiment orientation expressed toward a given target, i.e. an entity or its attributes (commonly referred to as *aspects*). Although entities (like products, services, topics, issues, persons, organizations or events) and their aspects can be organized in a hierarchy of parts and sub-parts as nested nodes [7, 11] following the *part-of* relationships, most of the research in opinion mining/retrieval neglects such complex organization of concepts, and prefers a simpler model where the target of an opinion is generically an aspect, which denotes both parts and attributes. However, during a decision-making process many aspects at different levels of granularity can be involved.

For example, booking a room usually requires the matching of different criteria on a subjective base. *Cleanliness* and *view* can be considered as two sub-aspects of the general concept of *room*, which along with *location* represent two aspects of the entity *hotel* (Fig. 2).

However, most of the existing systems merely present a flat list of aspects. Such lists are predefined and manually created, they usually reflect broad coverage aspects,

Fig. 2 Entity/aspects/sub-aspects hierarchy

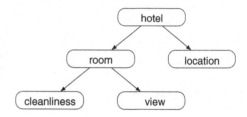

of which there is no guarantee of occurrence in the opinionated text. This is due to the fact that aspects are not extracted from text—then they not reflect the text content—but rather summarize, as a graded scale, the general opinion of the reviewer in few common points. Moreover, the hierarchical organization of aspects, which follows the *part-of* relationship, strictly depends on the target domain. *Soundtrack* is one of many aspects related to a *film*, but can be considered as an entity itself if we draw up the best top 100 soundtracks ever in the *music* domain.

This chapter proposes SABRE, an opinion retrieval system able to:

1. Extract from a given text aspects and their potential sub-aspects.
2. Associate to each aspect the corresponding opinion.
3. Detect the sentiment (positive or negative) of each opinion.
4. Retrieve documents which express an opinion about a given query.

A core component of SABRE is the aspect extraction one, which automatically extracts from text the hierarchy of aspects related to a given entity. However, in order to simplify the problem, the algorithm uses only the nodes at first level of the hierarchy (aspects) and considers all the sibling of this level as sub-aspects.

SABRE exploits such information for: (1) Re-ranking, during the second stage of the opinion retrieval, when the sentiment associated to each pair is exploited in combination to the relevance score obtained from the retrieval model; (2) Filtering, in order to improve the visualisation of reviews and help the user to filter out non relevant information during the navigation of the results.

To enable these operations, given Σ the set of available aspects, and a document $D = (p_1, p_2, \ldots, p_n)$ split up in text units , SABRE extracts a set of quintuples in the form of $(p_i, a_{ij}, a_{ijk}, s_{ijk}^{rel}, s_{ijk}^{sent})$, where:

- p_i is the text unit, it can be anyone of the possible ways of splitting a document, like sentences, paragraphs, or sliding windows;
- a_{ij} are the main aspects, $a_{ij} \in \Sigma$;
- a_{ijk} are the sub-aspect of a_{ij}, $a_{ijk} \in \Sigma \cup \{*\}$, with $*$ denoting the absence of sub-aspects;
- s_{ijk}^{rel} is the relevance weight of the couple $\langle a_{ij}, a_{ijk} \rangle$ within the text unit p_i, $s_{ijk}^{rel} \geq 0$;
- s_{ijk}^{sent} is the sentiment weight associated to $\langle a_{ij}, a_{ijk} \rangle$, it represent the polarity of the opinion expressed on that given pair, $s_{ijk}^{sent} \in [-1, 1]$.

The symbol $*$ denotes the lack of sub-aspects. Although a hierarchy defines the relationship between aspects and sub-aspects, the presence of a $\langle aspect, sub\text{-}aspect \rangle$ in a quintuple does not imply by default the existence of $\langle aspect, * \rangle$; i.e. $(p_i, a_{ij}, a_{ijk}, \cdot, \cdot) \not\Rightarrow (p_i, a_{ij}, *, \cdot, \cdot)$. Several quintuples can be associated to the same text unit, representing in this way the possibility of different (and maybe contrasting) opinions on the same aspect/sub-aspect, like in the sentence "the hotel was clean, but quite noisy". Moreover, such a definition makes the retrieval of opinions on a target entity/aspect/sub-aspect easier, with the possibility of expressing constraints on s_{ijk}^{sent}, the polarity of the opinion.

3   Aspect Extraction

There are two main categories of aspect extraction algorithms: the *frequency-based*, which rely on statistical analysis of corpora, and the *topic modeling*, which make use of more sophisticated machine learning approaches. This work exploits two frequency-based approaches: a baseline method based on term probabilities, and a model that grasps the different use of language between a specific domain and a general context. Both these algorithm rely on the simple observation that aspects and sub-aspects frequently occur as nouns. On this assumption, we built the two different methods described below.

3.1   BASE: A Simple Frequency-Based Algorithm

Frequency-based approaches compute statistics on term distributions from a training set, whose quality drives the effectiveness of the algorithm; when a new document come in, aspects are extracted on the base of their previously computed distributions.

The BASE algorithm, that will be used as a baseline algorithm, initially extracts from the training set of documents all the occurrences and co-occurrences of noun terms in a given sliding window s. During this process, the algorithm removes the stop-words, or the most k frequent terms, in order to avoid too frequent and non informative words. Then, given the set of extracted terms $T = (t_1, t_2, \ldots, t_m)$, the algorithm computes (1) the probability of the term t_i appearing as a noun in the sliding window and (2) the probability of a term t_j appearing as a noun in the sliding window given the occurrence of the term t_i as follows:

$$P(t_i) = \frac{freq(t_i)}{\sum_i freq(t_i)}. \tag{1}$$

$$P(t_j|t_i) = \frac{freq(t_i, t_j)}{freq(t_i)}. \tag{2}$$

Algorithm 1 shows the main steps. The list of extracted nouns represents the set of main aspects in the form $\langle a_{ij}, * \rangle$ with the associated relevance weight s_{ijk}^{rel} given by $P(a_{ij})$.

Then, the list of pairs $\langle a_{ij}, a_{ijk} \rangle$ is built weighting each $\langle aspect, sub\text{-}aspect \rangle$ accordingly to the relevance weight given by $P(a_{ijk}|a_{ij})$. However, in this second step the algorithm takes into account only those terms co-occurring with the N most frequent terms in the whole dataset (line 9). Then the algorithm keeps only the top z aspects weighed accordingly to their relevance scores. We exploit the output of this algorithm as baseline.

Algorithm 1 Aspect Extraction Baseline

Require: Unit text T_i, threshold z
Ensure: List A of pairs $\langle aspect, sub\text{-}aspect \rangle$ with associated weight s_{ijk}^{rel}

1: $A \leftarrow newList()$
2: $N \leftarrow nouns(T_i)$
3: $NC \leftarrow nounCoOccurrences(T)$
4: **for all** $t_j \in N$ **do**
5:     $s_{ij*}^{rel} \leftarrow P(t_j)$
6:     add $\langle t_j, * \rangle$ to A
7: **end for**
8: **for all** $\langle t_j, t_k \rangle \in NC$ **do**
9:     **if** $t_k \in mostCommonNouns()$ **then**
10:         $s_{ijk}^{rel} \leftarrow P(t_k|t_j)$
11:         add $\langle t_j, t_k \rangle$ to A
12:     **end if**
13: **end for**
14: sort A by s_{ijk}^{rel}
15: keep first z elements of A
16: **return** A

3.2 LM: Measuring the Divergence Between Languages

This algorithm is based on the idea that language differs when talking about a specific domain with respect to a general topic; then, this method aims at selecting the aspects whose distributions in a specific domain diverge from those in a general corpus, like the British National Corpus[1] (BNC).

To this extent, we exploit the Kullback-Leibler divergence (KL-divergence), a non-symmetric measure of the difference between two distributions. The KL-divergence measures the relevance of a term with respect to the difference between two distributions—one computed on the specific domain while the other on a generic corpus—as the information that the term conveys.

---

[1]http://www.natcorp.ox.ac.uk/.

However, to compute such a difference in a specific point, we make use of the pointwise Kullback-Leibler divergence, defined as follows:

$$\delta_t(p\|q) = p(t)log\frac{p(t)}{q(t)}. \tag{3}$$

where p is the distribution over the domain corpus and q is the distribution over the general corpus. Differently from the KL-divergence, the pointwise KL-divergence can assume negative values of δ, which correspond to non relevant aspects. However, in order to build the list of main aspects for a given text, we consider all noun terms t with $\delta_t(p\|q) > \varepsilon$ ($\varepsilon \geq 0$ threshold). Let denote with P_{domain} and $P_{general}$ the two distributions of a term on a domain and a general corpora. The term t can be considered as a main aspect if

$$\delta_t(P_{domain}(t)\|P_{general}(t)) > \varepsilon. \tag{4}$$

The threshold ε impacts the relevance of aspects in the domain corpus. However, the method still works for $\varepsilon = 0$, since in that case all non relevant aspects will take on $\delta_t < 0$. Another interesting point is that δ induces an order relation on the set of aspects: given two aspects a_1 and a_2, a_1 is more relevant than a_2 in the given domain if and only if $\delta_{a_1} > \delta_{a_2}$. The main steps of this method are showed in Algorithm 2.

Algorithm 2 LM Main Aspect Extraction

Require: Unit text T_i, threshold ε
Ensure: List A of main aspect with associated weight s_{ij*}^{rel}

1: $A \leftarrow newList()$
2: $N \leftarrow nouns(T)$
3: **for all** $t_j \in N$ **do**
4:     **if** $\delta_{t_j}(P_{domain}(t_j)\|P_{general}(t_j)) > \varepsilon$ **then**
5:         $s_{ij*}^{rel} \leftarrow \delta_{t_j}$
6:         add $\langle t_j, \cdot \rangle$ to A
7:     **end if**
8: **end for**
9: **return** A

3.2.1 Sub-aspect Extraction

The output of Algorithm 2 is a list of main aspects that represents the input to the algorithm for sub-aspect extraction. This phase exploits two measure of "quality" [17] of a sub-aspect defined as:

- **Phraseness**, the information lost following the adoption of a unigram (LM^1) in place of n-gram (LM^N) language model:

$$\varphi_{ph} = \delta_t(LM^N_{fg} \| LM^1_{fg}) ;$$ (5)

- **Informativeness**, the information lost when assuming that t is drawn from LM_{bg}—the *background* or general—rather than LM_{fg}—the *foreground* or domain—language model:

$$\varphi_i = \delta_t(LM^N_{fg} \| LM^N_{bg}).$$ (6)

The phraseness measures the information lost when words are considered as independent in a unigram language model rather than as a sequence in a n-gram model. The informativeness measures the information lost when assuming that as sentence has been drawn from the background (general) rather than the foreground (domain) corpus, in the case in point this measures the information added from the domain to the terms.

Since the quality of the pair $\langle aspect, sub\text{-}aspect \rangle$ is conditioned by both these factors, we define the relevance weight of the pair as:

$$\varphi = (\varphi_{ph} + \varphi_i) \times \mathcal{N}(\sigma^2).$$ (7)

$\mathcal{N}(\sigma^2)$ is used for smoothing the phraseness and informativeness weights, which are strongly regulated by words with high pointwise KL-divergence. $\mathcal{N}(\sigma^2)$ replaces the variance-to-mean ratio (VMR), since we observed that the distribution of co-occurrences follows a normal distribution.

For each main aspect extracted accordingly to (3), Algorithm 3 computes the relevance score φ for the pairs $\langle t_{ij}, t_{ijk} \rangle$, where t_{ijk} is a noun extracted from the text fragment. The final list of $\langle aspect, sub\text{-}aspect \rangle$ consists in all the pairs for which $\varphi > \varepsilon$.

Algorithm 3 LM Sub-Aspect Extraction

Require: Unit text T_i, main aspect list A, threshold ε
Ensure: List A of pairs $\langle aspect, sub\text{-}aspect \rangle$ with associated weight s^{rel}_{ijk}

1: $A \leftarrow newList()$
2: $N \leftarrow nouns(T_i)$
3: **for all** $t_j \in N$ **do**
4:     **for all** $t_k \in N$ **do**
5:         compute φ for $\langle t_j, t_k \rangle$
6:         $\varphi \leftarrow (\varphi_p + \varphi_i) \times \mathcal{N}(\sigma^2)$
7:         $s^{rel}_{ijk} \leftarrow \varphi$
8:         add $\langle t_j, t_k \rangle$ to A if $\varphi > \varepsilon$
9:     **end for**
10: **end for**
11: **return** A

3.3   Algorithm Extensions

Both frequency-based and language model divergence algorithms can be extended to consider phrases rather than terms. Specifically, we defined two possible extensions, which can be used either individually or combined:

Named entity recognition (NER):   This module is based on conditional random fields [5] in order to extract sequences of words which correspond to three types of named entity: person, organization, and location. This model has been trained on the CoNLL 2003[2] dataset.

Collocation (CL):   This module recognizes sequences of words that appear in the list of 50,000 bi-grams extracted from WordNet.

3.4   Sentiment Analysis

The algorithm used for assigning a sentiment score (s_{ijk}^{sent}) to each ⟨aspect, sub-aspect⟩ pair is a lexicon-based model that exploits the AFINN wordlist [12]. AFINN contains about 2500 English words that have been manually tagged with a score that can range from positive ($+5$) to negative (-5). The wordlist contains also some collocations while it disregards words with a neutral sentiment (score equal to zero).

For each text fragment, the algorithm computes the mean of sentiment scores associated to words in the text that appear in AFINN wordlist. However, when a "negation" word—like not, but, no, never, less, barely, hardly, rarely, aren't, weren't, won't, don't and isn't—is encountered, this reverses the sentiment score of all the words appearing at most at a distance of five terms from the negation. The score is rescaled in the interval $[-1, 1]$, and then associated to each ⟨aspect, sub-aspect⟩ extracted from the fragment.

4   Opinion Retrieval

The opinion retrieval engine is based on a two-stage approach:

1. A classical *tf-idf* vector space model [14] is employed for retrieving the top N documents ranked accordingly to the query terms.
2. The opinion ranking module re-ranks the top N documents in order to reflect their opinion scores.

While the document relevance in the first step is computed as in a standard vector space model (*tf-idf* weighing schema), the second step exploits the collection of

---

[2]http://www.cnts.ua.ac.be/conll2003/ner/.

quintuples $(p_j, a_{ij}, a_{ijk}, s_{ijk}^{rel}, s_{ijk}^{sent})$ associated to each text unit, and specifically the relevance (s^{rel}) and sentiment (s^{sent}) weights of each pair $\langle aspect, sub\text{-}aspect \rangle$.

Let $S = \{s \mid s = \langle s^{rel}, s^{sent} \rangle, \; s^{rel}$ and s^{sent} relevance and sentiment score of$\langle a_{ij}, a_{ijk} \rangle\}$ be the set relevance and sentiment score pairs extracted for all the $\langle aspect, sub\text{-}aspect \rangle$ in a document, the opinion score is calculated as:

$$\frac{\sum_{s \in S} |s^{rel} \cdot s^{sent}|}{|S|} \tag{8}$$

The opinion score has the advantages of taking into account both the relevance of the opinion and its polarity, in addition it normalizes this value with respect to number of $\langle aspect, sub\text{-}aspect \rangle$ pairs in a document.

Figure 3 shows the result set for the query "location breakfast" performed on a set of hotel reviews collected from TripAdvisor. Aspects extracted from the result set are listed on the left side of the main result list. Aspects are aggregated, on the basis of the pair $\langle aspect, sub\text{-}aspect \rangle$ they belong to, with their sentiment scores computed on the whole set of retrieved documents. Duplicate aspects denote the presence of the same aspect belonging to different pairs $\langle aspect, sub\text{-}aspect \rangle$.

Moreover, a filter based on the extracted aspects can be applied on the result set through the "aspect filter" button, which opens the window showed in Fig. 4.

Fig. 3 SABRE result page

☐ hotel (22.61)	☐ room (12.00)	☐ location (2.00)	☐ breakfast (1.36)
☐ location (0.0485)	☐ size (0.0491)	☐ value (0.2184)	☐ buffet (0.4487)
☐ review (0.0344)	☐ view (0.0435)	☐ hotel (0.1781)	☐ morning (0.1439)
☐ night (0.0325)	☐ bathroom	☐ everything	☐ value (0.0581)
☐ stay (0.0201)	(0.0172)	(0.1314)	☐ lunch (0.0369)
☐ star (0.0184)	☐ service (0.0148)	☐ great (0.0232)	☐ room (0.0262)
☐ staff (0.0125)	☐ hotel (0.0100)	☐ minute (0.0176)	☐ euro (0.0233)
☐ room (0.0108)	☐ club (0.0098)	☐ room (0.0138)	☐ coffee (0.0225)
☐ block (0.0079)	☐ noise (0.0084)	☐ walk (0.0124)	☐ selection
☐ family (0.0068)	☐ bath (0.0078)	☐ night (0.0091)	(0.0187)
☐ airport (0.0040)	☐ breakfast	☐ stay (0.0080)	☐ dinner (0.0180)
	(0.0073)	☐ tourist (0.0073)	☐ fruit (0.0150)
	☐ window (0.0067)		

Fig. 4 Aspect filter

5 Evaluation

SABRE has been evaluated in two different ways with respect to its ability to extract aspect/sub-aspect pairs. Each evaluation is performed on a dataset of 164,780 reviews from TripAdvisor, where each review has been anonymized. The TripAdvisor dataset represents the specific domain corpus. As global domain corpus we exploit the British National Corpus (BNC), which consist of about 4000 documents with 100 million words from different domains. The threshold z for the baseline is set to 30 aspects, while the threshold ε for aspect and sub-aspect in the LM approach is set to 10^{-3}; all the thresholds have been chosen after an empirical tuning of the system. The text has been analysed before the extraction of term distribution. The analysis comprises the tokenization, lemmatisation, stop-word removal, Part-Of-Speech tagging. Moreover, this pipeline includes also the named entity recognition and the collocation finding in the case of the two extensions explained in Sect. 3.3. Most of the text operations are performed by the Stanford CoreNLP API[3] [9], while the implementation of indexing and retrieval is performed on the top of ElasticSearch[4] engine.

---

[3] http://nlp.stanford.edu/software/corenlp.shtml.
[4] https://www.elastic.co/products/elasticsearch.

Table 1 Aspect labelling evaluation: (P)recision, (R)recall, (F)-measure

	⟨aspect, ∗⟩			⟨aspect, sub-aspect⟩		
	P	R	F	P	R	F
BASE	0.256	0.869	**0.396**	0.092	0.526	0.154
BASE-CF	0.254	0.869	0.393	0.093	0.530	0.155
BASE-NER	0.251	0.868	0.389	0.093	0.527	0.154
BASE-CF-NER	0.255	0.870	0.394	0.094	0.533	**0.157**
LM	0.279	0.873	0.422	0.148	0.448	0.211
LM + CF	0.281	0.878	**0.456**	0.148	0.453	0.211
LM + NER	0.278	0.873	0.421	0.144	0.448	0.207
LM + CF + NER	0.274	0.878	0.418	0.153	0.465	**0.230**

We report in bold the best F-measure values for both the baseline and the language modeling systems in the two different experiments: ⟨aspect, ∗⟩ and ⟨aspect, sub-aspect⟩ identification

5.1 Aspect Labelling

The first evaluation method is based on a manually labelled dataset built on a random selection of 200 out of 164,780 hotel reviews from TripAdvisor. The remaining 164,580 reviews were used for training the model. The annotator had to specify a pair ⟨aspect, sub-aspect⟩ for each review in the test set. We compared our extraction algorithm (**LM**) against the baseline (**BASE**) testing several configurations with and without the use of the collocation (**CF**) and named entity recognition (**NER**) extensions.

Table 1 reports the results of the evaluation when only the main aspect is considered, i.e. reducing all the labelled pairs to ⟨aspect, ∗⟩, and when the proper ⟨aspect, sub-aspect⟩ is identified. As expected, figures for ⟨aspect, sub-aspect⟩ identification are lower than those for detecting the main aspect, this is due to the more complex task, that here asks for the identification of a hierarchy between the aspects mentioned in the review.

In both experiments, the language model system achieves better performance than the baseline, however is only in the ⟨aspect, sub-aspect⟩ identification that the configuration with CF+NER gives the best result. However, it is important to underline here that the reported values should be considered only as a lower bound due to: (1) the small number of review considered, (2) the manual labelling performed by just one user, and (3) the inherent subjectivity in assessing the ⟨aspect, sub-aspect⟩ pairs.

5.2 User Feedback

Given the list of ⟨aspect, sub-aspect⟩ pairs extracted during the aspect labelling evaluation from the best system (LM + CF + NER), we asked 61 users to manually tag a sub-set of the pairs extracted from 97 reviews as relevant/not-relevant with

Table 2 User feedback evaluation: (P)recision, (R)recall, (F)-measure

	P	R	F
$\langle aspect, * \rangle$	0.416	0.933	0.547
$\langle aspect, sub\text{-}aspect \rangle$	0.232	0.879	0.351

respect to the review. The evaluation aims at finding out the number of $\langle aspect, sub\text{-}aspect \rangle$ pairs the user find as prominent for the given review.

The assessment took place in two-steps:

1. The user selects the main aspects from the list. Each user is given from 3 to 6 aspects from which she/he has to select those relevant for the given text unit.
2. A list of sub-aspects is generated from aspects selected at the previous step, among these the user chooses those more relevant for the given main aspect and text unit.

Table 2 reports the result of this evaluation. The evaluation shows high values of recall, these figures are expected since the labelling is performed on the list of predefined aspects returned by the algorithm. More interesting in this context are the precision values, which are higher than those reported in Table 1, and similarly to the previous experiment we notice a drop in performance when the $\langle aspect, sub\text{-}aspect \rangle$ has to be identified.

6 Related Work

The problem of opinion retrieval with respect to specific aspects/sub-aspects of interest is quite new, and to the best of our knowledge it has still to be addressed. However, if we consider each problem on its own, i.e. opinion retrieval and aspect-based opinion mining, they are two well rooted problems in their respective fields.

The opinion retrieval (OR) has been treated as an extension of the information retrieval (IR) task. Usually, OR is performed in two-stages. First a set of relevant document is retrieved, and then this set is re-ranked according to their opinion scores [7, Chap. 9] adopting machine learning or lexicon based approaches, this is also the approach adopted in SABRE. Most of the research on OR has been conducted within the TREC Blog Track evaluations, and all best systems participating in the opinion finding task of the three Blog Task evaluation (2006, 2007, 2008) followed such kind of strategy [6, 20, 21]. However, exception exists like the system proposed by Zhang et al. [19], where the two components are merged altogether.

Although we applied a simple lexicon-based approach, more sophisticated techniques have been developed to classify a sentence with respect to the sentiment it expresses. In addition to techniques based on the presence of some sentiment words [3], many methods are based on some machine learning techniques, both supervised [10, 13] and unsupervised [18].

Topic modelling is one of the main approaches adopted for aspect extraction. These methods are usually based on latent Dirichlet allocation (LDA) [2] or probabilistic latent semantic analysis (pLSA) [4], which are statistical methods for detecting the topics of a discussion. Then, they are exploited in the aspect extraction where each topic is an aspect. However, since these techniques try to capture the different distributions of terms in documents that treat different topics, their use in review domain is a bit tricky, due to fact that reviews tend to treat always the same topics. Titov and McDonald [16] propose a system based on two levels: the first one uses LDA for entity extraction, while the second extracts aspects considering only the neighbour of the given entity, neglecting the possibility of more level of aspect organization. Although some extension of LDA have been exploited to derive hierarchy [1, 8, 15], these methods are still too complex and require big training data and parameter tuning.

7 Conclusions

This chapter introduced SABRE, a two-stage aspect-based opinion retrieval system which takes into account hierarchy of aspects organized at two levels. We described the general system architecture, and explained how the information about aspects and sub-aspects is exploited for computing the opinion score during the re-rank, and for filtering during the navigation of the relevant documents.

The core of our system relies on the aspect extraction algorithm. We proposed to chose candidate terms exploiting the Kullback-Leibler divergence from a domain and a general purpose corpora. At an early stage, we conducted an evaluation to assess the capability of the proposed algorithm at extracting good candidate terms as aspects and sub-aspects. The evaluation demonstrated competitive results with respect to the baseline.

Most of current datasets for opinion retrieval rely on either TREC Blog Track or Twitter retrieval. None of them specifically focuses on aspect hierarchy extracted from the text. We plan to design a opinion retrieval evaluation that would benefit from such organization. As future work, we plan a thoroughly investigation for assessing the retrieval performance of SABRE.

References

1. Blei, D.M., Griffiths, T.L., Jordan, M.I.: The nested chinese restaurant process and bayesian nonparametric inference of topic hierarchies. J. ACM **57**(2), 7:1–7:30 (2010)
2. Blei, D.M., Ng, A.Y., Jordan, M.I.: Latent Dirichlet allocation. J. Mach. Learn. Res. **3**, 993–1022 (2003)
3. Godbole, N., Srinivasaiah, M., Skiena, S.: Large-scale sentiment analysis for news and blogs. In: Glance, N.S., Nicolov, N., Adar, E., Hurst, M., Liberman, M., Salvetti, F. (eds.) Proceedings of the First International Conference on Weblogs and Social Media, ICWSM 2007, Boulder, Colorado, USA, 26–28 March 2007

4. Hofmann, T.: Probabilistic latent semantic indexing. In: Proceedings of the 22nd Annual International ACM SIGIR Conference on Research and Development in Information Retrieval. SIGIR '99, pp. 50–57. ACM, New York, NY, USA (1999)
5. Lafferty, J.D., McCallum, A., Pereira, F.C.N.: Conditional random fields: probabilistic models for segmenting and labeling sequence data. In: Brodley, C.E., Danyluk, A.P. (eds.) Proceedings of the Eighteenth International Conference on Machine Learning (ICML 2001), Williams College, Williamstown, MA, USA, pp. 282–289. Morgan Kaufmann, 28 June–1 July 2001
6. Lee, Y., Na, S., Kim, J., Nam, S., Jung, H., Lee, J.: KLE at TREC 2008 blog track: blog post and feed retrieval. In: Voorhees, E.M., Buckland, L.P. (eds.) Proceedings of the Seventeenth Text REtrieval Conference, TREC 2008, Gaithersburg, MD, USA, 18–21 November 2008, vol. Special Publication 500-277. National Institute of Standards and Technology (NIST) (2008)
7. Liu, B.: Sentiment analysis and opinion mining. Synth. Lect. Hum. Lang. Technol. 5(1), 1–167 (2012)
8. Lu, B., Ott, M., Cardie, C., Tsou, B.K.: Multi-aspect sentiment analysis with topic models. In: Proceedings of the 2011 IEEE 11th International Conference on Data Mining Workshops. ICDMW '11, pp. 81–88. IEEE Computer Society, Washington, DC, USA (2011)
9. Manning, C.D., Surdeanu, M., Bauer, J., Finkel, J., Bethard, S.J., McClosky, D.: The stanford CoreNLP natural language processing toolkit. In: Proceedings of 52nd Annual Meeting of the Association for Computational Linguistics: System Demonstrations, pp. 55–60 (2014)
10. Mei, Q., Ling, X., Wondra, M., Su, H., Zhai, C.: Topic sentiment mixture: modeling facets and opinions in weblogs. In: Proceedings of the 16th International Conference on World Wide Web. WWW '07, pp. 171–180. ACM, New York, NY, USA (2007)
11. Moghaddam, S., Ester, M.: Aspect-based opinion mining from product reviews. In: Proceedings of the 35th International ACM SIGIR Conference on Research and Development in Information Retrieval. SIGIR '12, pp. 1184–1184. ACM, New York, NY, USA (2012)
12. Nielsen, F.Å.: A new anew: Evaluation of a word list for sentiment analysis in microblogs. In: Rowe, M., Stankovic, M., Dadzie, A., Hardey, M. (eds.) Proceedings of the ESWC2011 Workshop on 'Making Sense of Microposts': Big Things Come in Small Packages, Heraklion, Crete, Greece, 30 May 2011, CEUR Workshop Proceedings, vol. 718, pp. 93–98. CEUR-WS.org (2011)
13. Pang, B., Lee, L., Vaithyanathan, S.: Thumbs up?: Sentiment classification using machine learning techniques. In: Proceedings of the ACL-02 Conference on Empirical Methods in Natural Language Processing. EMNLP '02, vol. 10, pp. 79–86. Association for Computational Linguistics, Stroudsburg, PA, USA (2002)
14. Salton, G.: The SMART Retrieval System: Experiments in Automatic Document Processing. Prentice-Hall, Upper Saddle River (1971)
15. Teh, Y.W., Jordan, M.I., Beal, M.J., Blei, D.M.: Hierarchical dirichlet processes. J. Am. Stat. Assoc. 101(476), 1566–1581 (2006)
16. Titov, I., McDonald, R.T.: A joint model of text and aspect ratings for sentiment summarization. In: McKeown, K., Moore, J.D., Teufel, S., Allan, J., Furui, S. (eds.) ACL 2008, Proceedings of the 46th Annual Meeting of the Association for Computational Linguistics, 15–20 June 2008, pp. 308–316. Columbus, OH, USA (2008)
17. Tomokiyo, T., Hurst, M.: A language model approach to keyphrase extraction. In: Proceedings of the ACL 2003 Workshop on Multiword Expressions: Analysis. Acquisition and Treatment, vol. 18, MWE '03, pp. 33–40. Association for Computational Linguistics, Stroudsburg, PA, USA (2003)
18. Turney, P.D.: Thumbs up or thumbs down?: Semantic orientation applied to unsupervised classification of reviews. In: Proceedings of the 40th Annual Meeting on Association for Computational Linguistics. ACL '02, pp. 417–424. Association for Computational Linguistics, Stroudsburg, PA, USA (2002)
19. Zhang, W., Yu, C., Meng, W.: Opinion retrieval from blogs. In: Proceedings of the Sixteenth ACM Conference on Conference on Information and Knowledge Management. CIKM '07, pp. 831–840. ACM, New York, NY, USA (2007)

20. Zhang, W., Yu, C.T.: UIC at TREC 2006 blog track. In: Voorhees, E.M., Buckland, L.P. (eds.) Proceedings of the Fifteenth Text REtrieval Conference, TREC 2006, Gaithersburg, Maryland, 14–17 November 2006, vol. Special Publication 500-272. National Institute of Standards and Technology (NIST) (2006)
21. Zhang, W., Yu, C.T.: UIC at TREC 2007 blog track. In: Voorhees, E.M., Buckland, L.P. (eds.) Proceedings of The Sixteenth Text REtrieval Conference, TREC 2007, Gaithersburg, Maryland, USA, 5–9 November 2007, vol. Special Publication 500-274. National Institute of Standards and Technology (NIST) (2007)

Monitoring and Supporting People that Need Assistance: The BackHome Experience

Xavier Rafael-Palou, Eloisa Vargiu, Stefan Dauwalder
and Felip Miralles

Abstract People that need assistance, as for instance elderly or disabled people, may be affected by a decline in daily functioning that usually involves the reduction and discontinuity in daily routines and a worsening in the overall quality of life. Thus, there is the need to intelligent systems able to monitor indoor and outdoor activities of users to detect emergencies, recognize activities, send notifications, and provide a summary of all the relevant information. In this chapter, we present a sensor-based telemonitoring system that addresses all that issues. Its goal is twofold: (i) helping and supporting people (e.g. elderly or disabled) at home; and (ii) giving a feedback to therapists, caregivers, and relatives about the evolution of the status, behavior and habits of each monitored user. The proposed system is part of the EU project BackHome and it is currently running in three end-user's homes in Belfast. Our experience in applying the system to monitor and assist people with severe disabilities is illustrated.

1 Introduction

Decline in daily functioning usually involves the reduction and discontinuity in daily routines; entailing a considerable decrease of the quality of life (QoL). This is especially relevant for people that need assistance, as for instance elderly or disabled people [1]. Sometimes it may also hide pathological (e.g. Alzheimer) and/or mental (e.g. depression or melancholia) conditions.

X. Rafael-Palou (✉) · E. Vargiu · S. Dauwalder · F. Miralles
eHealth Unit, Eurecat, Barcelona, Spain
e-mail: xavier.rafael@eurecat.org

E. Vargiu
e-mail: eloisa.vargiu@eurecat.org

S. Dauwalder
e-mail: stefan.dauwalder@eurecat.org

F. Miralles
e-mail: felip.miralles@eurecat.org

© Springer International Publishing AG 2017
C. Lai et al. (eds.), *Information Filtering and Retrieval*,
Studies in Computational Intelligence 668, DOI 10.1007/978-3-319-46135-9_5

To remotely monitor and support this kind of people, especially those that live alone, novel and intelligent systems are required. In particular, ambient and assisting living systems must be deeply investigated to facilitate improving autonomy, safety and social participation of people with special needs, normally elder and/or disabled. Therefore, there is the need of systems that allow monitoring indoor and outdoor activities of users to detect emergencies, recognize activities, send notifications, and provide a summary of all the relevant information related to user daily activities. Ambient assisting living solutions normally use sensor-based solutions [11]. Through the sensors, a lot of information are gathered and suitable support given to the end-user, accordingly. In fact, once data have been analyzed, the system has to react and perform some actions. On the one hand, the user (e.g. elderly or disabled people) needs to be keep informed about emergencies as soon as they happen and s/he has to be in contact with therapists and caregivers to change habits and/or to perform some therapy. On the other hand, monitoring systems are very important from the perspective of therapists, caregivers, and relatives. In fact, those systems allow them to become aware of user context by acquiring heterogeneous data coming from sensors and other sources.

In this chapter, we present a novel solution that provides all the above-mentioned functionalities. In fact, it advances existing telemonitoring systems because it combines wireless off-the-shelf sensors, activity recognition, and anomaly/emergency detection providing a daily summary of the relevant performed activities in real time. This information is automatically shared with the therapists to allow decision making and defining personalized rules according to the user profile. The overall telemonitoring system is part of the BackHome project[1] and it has been developed according to a user-centered design approach in order to collect requirements and feedback from all the actors. The system was tested in two end-user's home in Belfast. The overall experience in using the system by people with severe disabilities is discussed in the chapter.

The rest of the chapter is organized as follows. Section 2 summarizes relevant related work in the field of telemonitoring focusing in particular on intelligent monitoring solutions. Section 3 illustrates the proposed solution presenting all the main components. In Sect. 4, the BackHome experience is presented focusing on all the implemented services and the corresponding results. Section 5 ends the chapter with conclusions.

---

[1]http://www.backhome-fp7.eu/.

2 Background

2.1 Telemonitoring

In the literature, various studies and systems aimed at monitoring and supporting people that need assistance, especially detecting and overwhelming the worsening in daily activities, have been proposed. Several methods are limited to measuring daily functioning using self-report such as with the modified Katz ADL scale [21] or a more-objective measurement method as the assessment of motor and process skills [8]. Recently, solutions have been proposed to unobtrusively monitor activities of people that need assistance. In particular, sensor-based approaches are normally used [15].

Sensor-based systems rely on a conjunction of sensors, each one devoted to monitor a specific status, a specific activity or activities related to a specific location. Binary sensors are currently the most adopted sensors [18], even if they are prone to noise and errors [17]. Once all of the data have been collected, intelligent solutions that incrementally and continuously analyze the data to all the involved actors (i.e. therapists, caregivers, relatives, and end-users themselves) are required. Moreover, it is then necessary to identify if the person needs a form of assistance since an unusual activity has been recognized. This requires the adoption of machine learning solutions to take into account the environment, the performed activity and/or some physiological data [4].

2.2 Activity Recognition in Telemonitoring Systems

There is a large literature on recognition of activities at home [19, 24]. At the same time, we find a great variability in the settings of the experiments either in the number of sensors and their type, individuals involved or the duration thereof. Also noteworthy is the large amount of recognition techniques (supervised, either generative or discriminative; and unsupervised).

A former study [18] already points out some of the difficulties in discriminating daily life activities based only on binary sensors activities. The automatic recognition system was based on rules defined from the context and the duration of the activities to identify. The data of the study were obtained from 14 days of monitoring activities at home. Although promising accuracies were achieved for some activities, detection tasks such as "leaving home" were nothing less than satisfactory with 0.2 of accuracy. This was because the activities were represented by rules directly defined on the firings outputted by single sensors (i.e. door switches); so they did not contemplate that could be activated for other reasons and in varying times, which made reduce their discriminating power.

A more exhaustive work regarding the use of switch and motion sensors for tracking people inside home is found in [23]. Tests were done with up to three simultaneous users. High performances were reported by the trained tracking models. However it is interesting to note that this type of sensors experimented occasional lag between "entering" a room and triggering a sensor; making to decrease the performance of the tracking models.

In [9] a more complex template learning model (SVM) was used to automatically recognize among 11 different home activities. The proposed technique was integrated in different sliding window strategies (e.g. weighting sensor events, dynamic window lengths, or two levels of window lengths). They used 6 months of data from three different homes in which activities such as "entering" or "leaving home" were monitored. From the best experimental settings the authors claimed accuracy for "entering" home about 0.80 of F1-score but around 0.4 for "leaving" home tasks.

In a more extensive work [3] they use Naïve Bayes (NB), Hidden Markov (HMM) models and conditional random fields (CRF) for the activity recognition problem. In this study, seven smart environments were used and 11 different data sets were obtained. Several activities were attempted to be recognized. Among others, we highlight "entering" and "leaving home" as relevant for our approach. Although they did not report specific accuracies for these activities, authors claimed an overall recognition performance on the combined dataset of 0.74 for the NB classifier, 0.75 for the HMM model, and 0.72 for the CRF using 3-fold cross validation over the set of annotated activities.

In [14], authors proposed a hybrid approach to recognize ADLs from home environments using a network of binary sensors. Among the different activities recognized "leaving" was one of them. The hybrid system proposed was composed by using an SVM to estimate the emission probabilities of an HMM. The results showed how the combination of discriminative and generative models is more accurate than either of the models on their own. Among the different schemes evaluated, the SVM/HMM hybrid approach obtains a significant 0.7 of F1-score a notable better performance than the rest of approaches.

Finally, detecting "multiple" people in single room by using binary sensors was already studied in an early work [22]. In that work, authors proposed a method based in expectation maximization Monte Carlo algorithm. In a more recent article [13], high accuracy (0.85) were reported on detecting visits at home using binary sensors. In that approach, they used an HMM algorithm over the room events although not all rooms of the home were monitored.

3   The Telemonitoring and Home Support System

To monitor users' activities, we develop a sensor-based telemonitoring and home support system (SB-TMHSS) able to monitor the evolution of the user's daily life activity. The implemented system is able to monitor indoor activities by relying on a set of home automation sensors and outdoor activities by using Moves.[2] Information

---

[2]http://www.moves-app.com/.

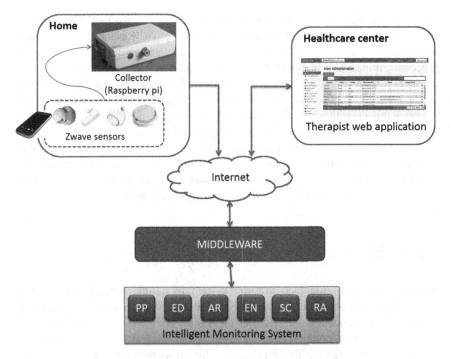

Fig. 1 Main components of the sensor-based system focused on the intelligent monitoring

gathered by the SB-TMHSS is also used to provide context-awareness by relying on ambient intelligence [2]. Monitoring users' activities through the SB-TMHSS gives us also the possibility to automatically assess quality of life of people [20].

The high-level architecture of the SB-TMHSS is depicted in Fig. 1. As shown, its main components are: home; middleware; intelligent monitoring system; and healthcare center.

3.1 Home

At home, a set of sensors are installed. In particular, we use presence sensors (i.e. Everspring SP103), to identify the room where the user is located (one sensor for each monitored room); a door sensor (i.e. Vision ZD 2012), to detect when the user enters or exits the premises; electrical power meters and switches, to control leisure activities (e.g. television and pc); and pressure mats (i.e. bed and seat sensors) to measure the time spent in bed (wheelchair). The system is also composed of a network of environmental sensors that measures and monitors environmental variables like temperature, but also potentially dangerous events like gas leak, fire, CO escape and presence of intruders. All the adopted sensors are wireless z-wave.[3] They send the retrieved data to a collector (based on Raspberry pi[4]). The Raspberry pi collects all the retrieved data and securely redirects them to the cloud where they will be stored, processed, mined, and analyzed. The proposed solution relies on z-wave technology for its efficiency, portability, interoperability, and commercial availability. In fact, on the contrary of other wireless solutions (e.g. ZigBee), z-wave sensors are able to communicate with any z-wave device. Moreover, we adopt a solution based on Raspberry pi because it is easy-to-use, cheap, and scalable.

We are also using the user's smartphone as a sensor by relying on Moves, an app for smartphones able to recognize physical activities and movements by transportation. Among the activity trackers currently on the market, we select Moves because it does not need user intervention being always active in background.

3.2 Middleware

As mentioned above, the telemonitoring system, by definition needs to be interoperable, extensible and scalable. In order to deal with such requirements the design of the system architecture was based on the service oriented architecture framework (SOA).[5] Thus, each component of the system was built as web service itself and communicate through the REST protocol. This protocol is selected since it works over HTTP/s and allows relaxed interoperability between different components. The middleware, which is the set of core components of the architecture, is composed by independent web services that provides not only scalability and interoperability but also securely interconnect the main components of the system, i.e. the home, the intelligent monitoring system, as well as the healthcare center. Its main functional components are:

- The API façade, which encapsulates a set of information services provided by the intelligent monitoring system in order to be consumed by external applications (i.e. the healthcare center). Internally, this component securely dispatches all requests of information from outside and redirects them to the enterprise service bus. This service contains policies and protocols to handle load balancing and concurrency.
- The enterprise service bus (ESB), which orchestrates the communication between internal components avoiding ad hoc communication. The ESB makes the system more extensible and flexible to changes by doing all components communicate through it.
- The security manager (SM), which provides mechanisms to user authentication, and services authorization. This component manages the session IDs to the system as well as the UUIDs that identify uniquely the users and the services which have access to.

---

[3]http://www.z-wave.com/.

[4]http://www.raspberrypi.org/.

[5]www.oasis-open.org/.

- Notification service (NS), which is a mechanism for sending asynchronously data (e.g. events, emergencies, rule actions) generated from the consumer (e.g. intelligent monitoring) to a receiver or client (e.g. health care center).

3.3   Intelligent Monitoring

In order to cope with data necessities of the actors of the system (i.e. therapists, caregivers, relatives, and end-users themselves), an intelligent monitoring (IM) system has been designed. It is aimed to continuously analyzing and mining the data through five-dimensions: detection of emergencies, activity recognition, event notifications, summary extraction, and rule triggering. In order to achieve these objectives, the intelligent monitoring system is composed of the following modules (see Fig. 2): PP, the pre-processing module to encode the data for the analysis; ED, the emergency detection module to notify, for instance, in case of smoke and gas leakage; AR, the activity recognition module to identify the location, position, activity- and sleeping-status of the user; EN, the event notification module to inform when a new event has been detected; SC, the summary computation module to perform summaries from the data; and RA, the risk advisement module to notify risks at runtime.

3.3.1   Pre-processing

IM continuously and concurrently listens for new data. The goal of PP is to preprocess the data iteratively sending a chunk c to ED and also RA according to a sliding window approach. Starting from the overall data streaming, the system sequentially considers a range of time $|t_i - t_{i+1}|$ between a sensor measure s_i at time

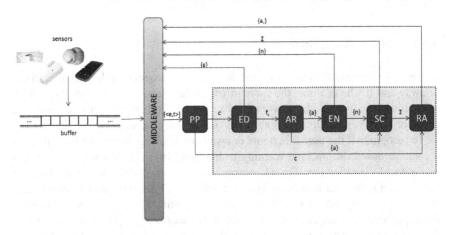

Fig. 2 The flow of data and interactions among the modules in the proposed approach

195	24.10	2014-02-24 10:21:54	195	24.10	2014-02-24 10:30:04	177	100	2014-02-24 10:31:55	195	24.10	2014-02-24 10:34:54

Fig. 3 Example of a chunk composed of four sensor measures

t_i and the subsequent measure s_{i+1} at time t_{i+1}. Thus, the output of PP is a window c from t_s to t_a, where t_s is the starting time of a given period (e.g. 8:00 a.m.) and t_a is the actual time. Thus, each chunk is composed of a sequence of sensor measures s; where s is a triple $\langle ID, v, t \rangle$, i.e. the sensor ID, its value and the time in which a change in the sensor status is measured. Figure 3 shows an example of a chunk composed by four sensors measures.

3.3.2 Emergency Detection

ED module aims to detect and inform about emergency situations for the end-users and sensor-based system critical failures. Regarding the critical situations for the end-users, an emergency is risen when specific values appear on c (e.g. gas sensor ID, smoke sensor ID). Regarding the system failures, ED is able to detect when the end-user's home is disconnected from the middleware as well as a malfunctioning of a sensor (e.g. low battery). The former is implemented by a keepalive mechanism in the Raspberry pi. If no signals are received from the Raspberry pi after a given threshold, an emergency is risen. The latter is implemented by using a multivariate gaussian distributions of sensor measurements on c. If the corresponding total number of measures is greater than a given threshold, an emergency is risen.

Each emergency is a pair $\langle s_i, l_{\varepsilon i} \rangle$ composed of the sensor measure s_i and the corresponding label $l_{\varepsilon i}$ that indicates the corresponding emergency (e.g. fire, smoke). Once the ED finishes the analysis of c, the list of emergencies ε is sent to the middleware, whereas c, filtered from the critical situations, is sent to AR.

3.3.3 Activity Recognition

In the current implementation, the system is able to recognize if the user is at home or away and if s/he is alone; the room in which the user is (no-room in case s/he is away, transition in case s/he moving from a room to another); the activity status (i.e. active or inactive); and the sleeping status (i.e. awake or asleep).

To recognize if the user is at home or away and if s/he is alone, we implemented a solution based on machine learning techniques [16]. The adopted solution is a hierarchical classifier (see Fig. 4) composed of two levels: the upper is aimed at recognizing if the user is at home or not, whereas the lower is aimed at recognizing if the user is really alone or if s/he received some visits. The goal of the classifier at the upper level is to improve performance of the door sensor. In fact, it may happen that the sensor registers a status change (from closed to open) even if the door has

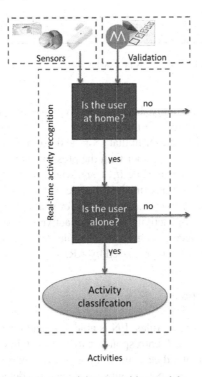

Fig. 4 The hierarchical approach in the activity recognition module

not been opened. This implies that AR may register that the user is away and, in the meanwhile, activities are detected at user's home. On the contrary, AR may register that the user is at home and, in the meanwhile, activities are not detected at user's home. Thus, we first revise the data gathered by AR searching for anomalies, i.e.: (1) the user is away and at home some events are detected and (2) the user is at home and no events are detected. Then, we validate those data by relying on Moves, installed and running on the user smartphone, and the supervision of the user. Using those as an "oracle", we build a dataset in which each entry is labeled depending on the fact that the door sensor was right (label "1") or wrong (label "0"). The goal of the classifier at the lower level is to identify whether the user is alone or not. The input data of this classifier are those that has been filtered by the upper level, being recognized as positives. To build this classifier, we rely on the novelty detection approach [10] used when data has few positive cases (i.e. anomalies) compared with the negatives (i.e. regular cases); in case of skewed data.

To measure the activity status, we rely on the home automation sensors. By default, we consider as "active" the status of the user when s/he is away (the corresponding positions are saved as "no-room"). On the contrary, when the user is at home, AR recognizes s/he as "inactive" if the sensor measures at time t_i that user is in a given room r and the following sensor measure is given at time t_{i+1} and the user was in the same room, with $t_{i+1} - t_i$ greater than a given threshold θ. Otherwise, the system classified the user as "active".

2014-02-24 10:21:54	2014-02-24 10:30:04	home, bedroom, inactive, asleep	2014-02-24 10:31:55	2014-02-24 10:34:54	home, bathroom, active, awake

Fig. 5 Example of a chunk after the AR processing

Finally, sleeping is currently detected by relying on the presence sensor located in the bedroom and the pressure mat located below the mattress. In particular, we consider the presence of the user in that room and no movements detection (i.e. the activity status is "inactive") together with the pressure of the mattress.

Thus, the output of AR is a triple $\langle t_s, t_e, l \rangle$, where t_s and t_e are the time in which the activity has started and has finished, respectively, and l is a list of four labels that indicates: the localization (i.e. home, away, or visits), the position (i.e. the room, no-room, or transition), the activity status (i.e. active or inactive), and the sleeping status (i.e. awake or asleep). To give an example, let us consider Fig. 5 where the same chunk of Fig. 3 has been processed by AR.

3.3.4 Event Notification

By relying on a set of simple rules, EN is able to detect events to be notified. Each event is defined by a pair $\langle t_i, l \rangle$ corresponding to the time t_i in which the event happens together with a label l that indicates the kind of event. In particular, according to user requirements and therapist's focus groups, we decided to detect the following kind of events: leaving the home, going back to home, receiving a visit, remaining alone after a visit, going to the bathroom, going out of the bathroom, going to sleep, and awaking. These events allow to study activity degradation as well as improvement/worsening of the overall quality of life. Nothing prevents to consider further notification and/or to change them in case requirements change or further needs arise. Following the example, in Fig. 3, an event is the pair $\langle$2014-02-24 10:31:55, *going to the bathroom*$\rangle$.

3.3.5 Summary Computation

Once all the activities and events have been classified, measures aimed at representing the summary of the user's monitoring during a given period are performed. In particular, two kinds of summary are provided: historical and actual. As for the historical summary, we decided to have a list of the activities performed during (i) the morning (i.e. from 8:00 a.m. to 8:00 p.m.), (ii) the night (i.e. from 8:00 p.m. to 8:00 a.m.), (iii) all the day, (iv) the week (from Monday morning to Sunday night), as well as (v) the month. In particular, we monitor: sleeping time; time spent outdoors; time spent indoors; time spent performing indoor activities; time spent performing outdoor activities; number of times spent in each room; and number of times that the user lefts the house. As for the actual summary, we are interested in monitoring: the room in which the user is; if the user is at home, or not; the number of times that s/he leaves the home; sleeping time; activity time; and number of visits per room.

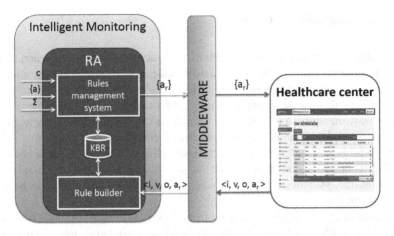

Fig. 6 The RA module functioning

As a final remark, let us note that all emergencies, activities, notifications, and summaries are stored in a database to be available to all the involved actors.

3.3.6 Risk Advisement

RA is aimed at advising therapists about one or more risky situations before they happen. The module executes the corresponding rules at runtime according to the sequence of sensor measures coming from the PP as well as the summary provided by the SC. Figure 6 sketches how RA works and which are its main components and interactions. Through the healthcare center, therapists access to an ad-hoc user interface to define the rules corresponding to risks. Those rules are automatically coded in a suitable language, namely ATML [7], and then translated in DRL by the Rule Builder and stored in the knowledge-based of rules (KBR). RA continuously processes data coming from the other modules of the IM and acts according the defined rules in KBR. In particular, it analyzes the entire chunk c from PP, the list of activities a from AR, and the complete summary Σ from SC. The actions $\{a_r\}$ triggered by RA are sent to the middleware that is in charge of actuate in consequence sending the corresponding advisement to the healthcare center.

To implement the RA we relied on Drools,[6] a rules management system that provides a rule execution server, and a web authoring and rules management application. A rule is a quadruple $\langle i, v, o, a_r \rangle$, where i is the item that has to be verified (e.g. a room, the number of slept hours) according to a given value v (e.g. bedroom, 4 slept hours); o is the logic operator (i.e. and, or, not) and a "null" operator in case there is only one term; and a is the action to be performed (i.e. send a notification, an alarm, or an email).

---

[6]http://www.drools.org/.

3.4 The Healthcare Center

The healthcare center receives notifications, summaries, statistics, and general information belonging to the users through a web application. Its goal is to keep informed therapists and caregivers about emergencies as soon as they happen and to proactively inform them about changes in user habits and/or to perform some therapy. In this way, therapists and caregivers become aware of user context by acquiring heterogeneous data coming from sensors and other sources.

The healthcare center has been implemented as a Web-based application and provides user management, rehabilitation task management, therapy assessment, rule definition, statistics on system usage, as well as for communication between therapists and user.

A modular approach was considered from the very starting point of the healthcare center design, which led to the definition of a loosely coupled system where each of its components keeps its logic as self-contained as possible [6]. This design strategy is crucial in order to manage changes while reducing its overall impact in the rest of the platform. As a result, each of the main functionality is encapsulated in a self-contained module, which in turn, is managed by the platform infrastructure services. Those base infrastructure services are in charge of the definition of a common application context where every module is registered while providing cross-platform functionality as well.

4 The BackHome Experience

The overall system presented in this chapter is part of the EU project BackHome.[7] The project is aimed at moving brain computer interfaces (BCIs) from being laboratory devices for healthy users toward practical devices used at home by people with limited mobility. This requires a system that is easy to set up, portable, and intuitive. Thus, BackHome aims to develop BCI systems into practical multimodal assistive technologies to provide useful solutions for communication, Web access, cognitive stimulation and environmental control, and to provide this technology for home usage with minimal support. These goals are attained through three key developments, each of them advancing the current state of the art [5, 12]: (i) practical electrodes, with the delivery of novel BCI equipment which sets a new standard of lightness, autonomy, comfort and reliability; (ii) easy-to-use software tailored to peoples needs, with a complete range of highly desirable applications finely tuned for one-click command and adaptive usage; and (iii) telemonitoring and home support to remotely assist independent use, with remote services to plan and monitor cognitive rehabilitation and pervasively assess the use of the system and the quality of life of the individual. The development followed a user-centered design approach in order to collect requirements and feedback from all the actors.

---

[7]http://www.backhome-fp7.eu/backhome/index.php.

A total of four participants were recruited in Belfast for a 6-week home based evaluation of the system. For the sake of anonymity let us refer to the final users as Home User 1 (HU1), Home User 2 (HU2), Home User 3 (HU3), and Home User 4 (HU4). Actually, only two of them concluded the 6-week evaluation period. In fact, attempts to set up the system in HU2's home failed due to the restricted Internet connection; whereas HU4 became ill after the installation of the BCI and did not recover in time to participate in the home based testing. Thus, the evaluation has been performed only with HU1 and HU3.

In each home, we installed a presence sensor in each room a door sensor in each entrance, and two power meter switches to control a light and a radio through the BCI. Environmental sensors have not been used in BackHome because of the user requirements, moreover, pressure mat sensors have not be installed due to privacy constraints inside the project.

Regarding the IM and due to the BackHome user requirements and end-user characteristics, in addition to PP (that is essential), AR, SC and RA was implemented.

AR was first evaluated with 2 able-bodied users in Barcelona, as reported in [16]. The hierarchical classifier showed an improvement of 15 % of accuracy with respect to a rule-based solution (see Table 1). To highlight the performance of the proposed approach, let us consider the Fig. 7 that shows a comparison between the real data, labeled during the validation phase (on the left), and the data classified by relying to the approach proposed in this chapter (on the right).

Table 1 Results of the overall hierarchical approach with respect to the rule-based one

Metric	Rule-based	Hierarchical	Improv. (%)
Accuracy	0.80	0.95	15
Precision	0.68	0.94	26
Recall	0.71	0.91	20
F_1	0.69	0.92	23

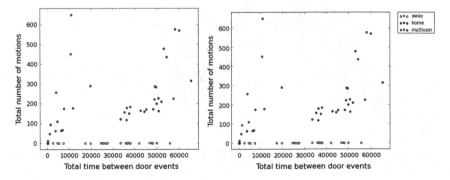

Fig. 7 Comparison between real labeled data and data classified by the hierarchical approach

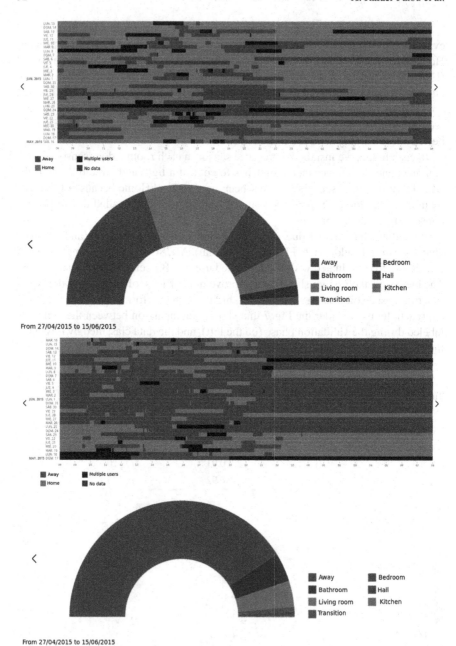

Fig. 8 Indoor user habits recognized by the SB-TMHSS for HU1 (the two graphs on the *top*) and HU3 (the two graphs on the *bottom*) for a period of 1 month

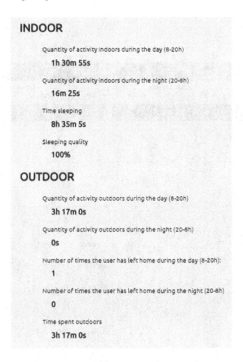

INDOOR

Quantity of activity indoors during the day (8-20h)
1h 30m 55s

Quantity of activity indoors during the night (20-8h)
16m 25s

Time sleeping
8h 35m 5s

Sleeping quality
100%

OUTDOOR

Quantity of activity outdoors during the day (8-20h)
3h 17m 0s

Quantity of activity outdoors during the night (20-8h)
0s

Number of times the user has left home during the day (8-20h):
1

Number of times the user has left home during the night (20-8h)
0

Time spent outdoors
3h 17m 0s

Fig. 9 An example of data coming from the sensors as shown to the therapist

Figure 8 shows the results belonging to 1 month of indoor monitoring; outdoor monitoring is reported similarly. In particular, for each user the following information is given: when the user was at home, away, or received visits day by day and a summary of the different locations in which the user was during a period of 1 month.

The Healthcare center in BackHome, called Therapist Station, allows therapists to remotely manage the end-user. Among the overall set of functionality provided by the therapist station, let us consider here the most relevant for monitoring and supporting.

Thanks to the IM, the Therapist Station daily receives the summary of the information regarding the daily-life activities of the user (computed by SC). Figure 9 shows an example of data coming from the sensors regarding both indoor and outdoor activities.

Moreover, therapists defined simple rules, such the one shown in Fig. 10. The therapist set a rule to raise an alarm if the user spends more than ten hours at bed in a day.

Overall the healthcare center was viewed in a positive light and considered to be an asset to daily practice. On average, the 36.63 % of the therapists evaluated as positive (4) the overall healthcare center and the 44.22 % as very positive (5), making a total of 80.86 % of positive and very positive evaluation. The Therapist Station was thought to be "modern and is relatively easy to use" and that the "site is laid out

Fig. 10 Trigger definition therapist interface

well". The focus of the BackHome therapist station was considered to be a "very useful starting point when a client returns home from hospital and is very dependent". Additionally, it might be useful to think of other populations and applications for the therapist station, "the Therapist Station is an excellent platform, could be used with a range of clients; in paediatrics would be good for assigning home programmes etc".

5   Conclusions

In this chapter, we presented a sensor-based system aimed at detecting emergencies, recognizing activities, sending notification as well as collecting the information in a summary and executing actions triggered by means of a rule-based engine. The goal of the implemented system was to monitor and support people that need assistance and to constantly give a suitable feedback to therapists and/or caregivers. They have access to information about the evolution of the status, behavior and habits of the corresponding user, thanks to a web-based application. Under the umbrella of BackHome, the system has been installed and tested in two disabled people's homes where is currently running and remotely tested by about 80 therapists.

The BackHome telemonitoring features have proven to be an effective way of remotely reporting information about end-user habits, quality of life and detailed usage of the BCI environment. Those telemonitoring capabilities of the BackHome system provide tools and means for therapists and technical experts to support end-users and caregivers at home in a reactive but also in a proactive way.

Acknowledgments The research leading to these results has received funding from the European Community's, Seventh Framework Programme FP7/2007-2013, BackHome project Grant Agreement No. 288566.

References

1. Bakkes, S., Morsch, R., Krose, B.: Telemonitoring for independently living elderly: inventory of needs and requirements. In: 2011 5th International Conference on Pervasive Computing Technologies for Healthcare (PervasiveHealth), pp. 152–159. IEEE (2011)
2. Casals, E., Cordero, J.A., Dauwalder, S., Fernández, J.M., Solà, M., Vargiu, E., Miralles, F.: Ambient intelligence by atml: rules in backhome. In: Lai, C., Giuliani A., Semeraro, G. (eds.) Emerging Ideas on Information Filtering and Retrieval. DART 2013: Revised and Invited Papers (2014)
3. Cook, D.J.: Learning setting-generalized activity models for smart spaces. IEEE Intell. Syst. **2010**(99), 1 (2010)
4. Cook, D.J., Augusto, J.C., Jakkula, V.R.: Ambient intelligence: technologies, applications, and opportunities (2007)
5. Edlinger, G., Hintermller, C., Halder, S., Vargiu, E., Miralles, F., Lowish, H., Anderson, N., Martin, S., Daly, J.: Brain neural computer interface for everyday home usage. In: HCI International 2015, pp. 437–446. Springer International Publishing (2015)
6. Fernández, J., Solà, M., Steblin, A., Vargiu, E., Miralles, F.: The relevance of providing useful information to therapists and caregivers in tele*. In: Lai, C., Giuliani, A., Semeraro, G. (eds.) DART 2014: Revised and Invited Papers (in press)
7. Fernández, J.M., Torrellas, S., Dauwalder, S., Solà, M., Vargui, E., Miralles, F.: Ambient-intelligence trigger markup language: a new approach to ambient intelligence rule definition. In: 13th Conference of the Italian Association for Artificial Intelligence (AI*IA 2013). CEUR Workshop Proceedings, vol. 1109 (2013)
8. Fisher, A.G., Jones, K.B.: Assessment of Motor and Process Skills. Three Star Press, Fort Collins (1999)
9. Krishnan, N.C., Cook, D.J.: Activity recognition on streaming sensor data. Pervas. Mobile Comput. **10**, 138–154 (2014)
10. Markou, M., Singh, S.: Novelty detection: a review? Part 1: statistical approaches. Signal Process. **83**(12), 2481–2497 (2003)
11. Meystre, S.: The current state of telemonitoring: a comment on the literature. Telemed J E Health **11**(1), 63–69 (2005)
12. Miralles, F., Vargiu, E., Dauwalder, S., Solà, M., Müller-Putz, G., Wriessnegger, S.C., Pinegger, A., Kübler, A., Halder, S., Käthner, I., Martin, S., Daly, J., Armstrong, E., Guger, C., Hintermüller, C., Lowish, H.: Brain computer interface on track to home. Sci. World J. Article ID 623896 (2015)

13. Nait Aicha, A., Englebienne, G., Kröse, B.: How lonely is your grandma?: Detecting the visits to assisted living elderly from wireless sensor network data. In: Proceedings of the 2013 ACM Conference on Pervasive and Ubiquitous Computing Adjunct Publication, pp. 1285–1294. ACM (2013)
14. Ordónez, F.J., de Toledo, P., Sanchis, A.: Activity recognition using hybrid generative/discriminative models on home environments using binary sensors. Sensors **13**(5), 5460–5477 (2013)
15. Pol, M.C., Poerbodipoero, S., Robben, S., Daams, J., Hartingsveldt, M., Vos, R., Rooij, S.E., Kröse, B., Buurman, B.M.: Sensor monitoring to measure and support daily functioning for independently living older people: a systematic review and road map for further development. J. Am. Geriatr. Soc. **61**(12), 2219–2227 (2013)
16. Rafael-Palou, X., Vargiu, E., Serra, G., Miralles, F.: Improving activity monitoring through a hierarchical approach. In: The International Conference on Information and Communication Technologies for Ageing Well and e-Health (ICT 4 Ageing Well) (2015)
17. Ranganathan, A., Al-Muhtadi, J., Campbell, R.H.: Reasoning about uncertain contexts in pervasive computing environments. IEEE Pervas. Comput. **3**(2), 62–70 (2004)
18. Tapia, E.M., Intille, S.S., Larson, K.: Activity Recognition in the Home Using Simple and Ubiquitous Sensors. Springer, Berlin (2004)
19. Van Kasteren, T., Noulas, A., Englebienne, G., Kröse, B.: Accurate activity recognition in a home setting. In: Proceedings of the 10th International Conference on Ubiquitous computing, pp. 1–9. ACM (2008)
20. Vargiu, E., Fernández, J.M., Miralles, F.: Context-aware based quality of life telemonitoring. In: Lai, C., Giuliani, A., Semeraro, G. (eds.) Distributed Systems and Applications of Information Filtering and Retrieval. DART 2012: Revised and Invited Papers (2014)
21. Weinberger, M., Samsa, G.P., Schmader, K., Greenberg, S.M., Carr, D., Wildman, D.: Comparing proxy and patients' perceptions of patients' functional status: results from an outpatient geriatric clinic. J. Am. Geriatr. Soc. **40**(6), 585–588 (1992)
22. Wilson, D., Atkeson, C.: Automatic health monitoring using anonymous, binary sensors. In: CHI Workshop on Keeping Elders Connected, pp. 1719–1720. Citeseer (2004)
23. Wilson, D.H., Atkeson, C.: Simultaneous tracking and activity recognition (STAR) using many anonymous, binary sensors. In: Pervasive Computing, pp. 62–79. Springer (2005)
24. Ye, J., Dobson, S., McKeever, S.: Situation identification techniques in pervasive computing: a review. Pervas. Mobile Comput. **8**(1), 36–66 (2012)

The Relevance of Providing Useful and Personalized Information to Therapists and Caregivers in Tele*

Juan Manuel Fernández, Marc Solà, Alexander Steblin,
Eloisa Vargiu and Felip Miralles

Abstract Nowadays, filtering and analyzing data coming from Tele* (i.e. telemedicine, telerehabiliation, telemonitoring, telecare, and teleassistance) systems is becoming more and more relevant. In fact, those systems gather a lot of data coming from patients through wearable, domotic, and environmental sensors, as well as questionnaires and interviews. The role of therapists and care givers is essential for remotely assisting the corresponding patients. Thus, intelligent solutions able to understand all those data and process them to keep therapists and caregivers aware about their assisted persons are needed. Moreover, friendly and useful tools for accessing and visualizing those data must be provided to therapists and caregivers. In this chapter, we present a generic Tele* solution that, in principle, may be customized to whatever kind of real scenarios to give a continuous and efficient support to therapists and caregivers. The aim of the proposed solution is to be as flexible as possible in order to be able to provide telerehabilitation, telemonitoring, teleassistance or a conjunction of them, depending on the real situation. Three customizations of the generic platform are also presented.

Marc Solà and Alexander Steblin contributed equally to this work.

J.M. Fernández (✉) · M. Solà · A. Steblin · E. Vargiu · F. Miralles
eHealth Unit, Eurecat, Barcelona, Spain
e-mail: juanmanuel.fernandez@eurecat.org

M. Solà
e-mail: marc.sola@eurecat.org

A. Steblinn
e-mail: alexander.steblin@eurecat.org

E. Vargiu
e-mail: eloisa.vargiu@eurecat.org

F. Miralles
e-mail: felip.miralles@eurecat.org

© Springer International Publishing AG 2017   97
C. Lai et al. (eds.), *Information Filtering and Retrieval*,
Studies in Computational Intelligence 668, DOI 10.1007/978-3-319-46135-9_6

1 Introduction

Thanks to the new and inexpensive technologies it is now possible and affordable to get information and realize medical tasks remotely. Telemedecine is now a reality and can be applied to several fields of the health practice to assist patients in self-care and adherence to treatments. Telerehabilitation enables following continuous interventions which may improve health conditions (both physical and cognitive) without the need for the patient to physically move to specialized facilities. Telemonitoring allows therapists and caregiver to be aware of the status of the patients. Teleassistance improves autonomy, safety and social participation of people with special needs, normally the elder and/or the disabled, through home support technologies which postpone the need of consumption of socio-sanitary services and associated costs.

Hereinafter, we refer to Tele* to name the group of remotely solutions in the field of the eHealth (i.e. Telemedicine, Telerehabilitation, Telemonitoring, and Teleassistance). Summarizing, Tele* allows: to improve the quality of clinical services, by facilitating the access to them, helping to break geographical barriers; to center the assistance in the patient, facilitating the communication between different clinical levels; to extend the therapeutic processes beyond the hospital, like patient's home; and a saving for unnecessary costs and a better costs/benefits ratio. The corresponding applications allow direct communication between the patient and the professional staff as a common denominator.

In this chapter, we present and discuss a generic Tele* solution, based on ICT, that has been used by real end-users in national and European projects. In particular, we present: (i) a home telerehabilitation solution for people with motor impairments; (ii) a mobile telerehabilitation and telemonitoring solution for people affected by chronic obstructive pulmonary disease (COPD); and (iii) a home telerehabilitation, telemonitoring, and teleassistance solution for people with severe disabilities. All the proposed solutions rely on a generic Tele* platform that is the focus of this chapter. Although, the end-user is the main actor of Tele*, a relevant role is played by therapists and caregivers that may access the gathered data receiving suitable notification and taking decisions accordingly. Thus, in this chapter, we face Tele* from the perspective of therapists and caregivers and the information they interact with.

The chapter is organized as follows, Sect. 2 introduces the problem and presents the definition of Tele* and some of the main application fields. In Sect. 3, we present our generic Tele* platform, whereas in Sect. 4, we illustrate how the generic Tele* platform has been customized in the three relevant real scenarios. Finally, Sect. 5 ends the chapter with some conclusions.

2 Background

Tele* platforms may include all kinds of medical devices that get information from the patient. Additionally, home sensor technology has created a new opportunity to remotely assist patients and to exploit the information gathered in this way [16]. Wearable sensors could also be used to provide more objective measures. The main advantage of wearable technology is the ability to measure motor behaviour under real-life conditions and for longer periods than could be observed in a clinical setting [25].

2.1 Telerehabilitation

The main objective of any rehabilitation program is to return the total, or the maximum, functionality after a health incident or illness. Telerehabilitation has the same aim adding also the possibility to follow a rehabilitation program scheduled by a therapist (e.g. at the patient's home). Telerehabilitation is becoming increasingly popular because it provides low cost, completely personalized therapies, with immediate feedback, quantifiable outcomes and has indicated significant therapeutic benefits [30].

Evidence to support the use of rehabilitation programs also for cognitive impairments has been growing for the last decade with examples extending to memory [9, 32], working memory [17, 20], attention [37], and visual perception [7, 19].

Independently of how we get the information, in a telerehabilitation system it is really important to process the information properly in order to provide a comprehensive assessment of daily functioning before, during and after the rehabilitation process. These systems allow to get the specific level of impairments or the effort limits that the user can do in a rehabilitation exercise. This information helps to determine the ability to perform the daily life activities.

2.2 Telemonitoring

We may define telemonitoring according to the Institute of Medicine in U.S. [13]: "the remote monitoring of patients including the use of audio, video, and other telecommunications and electronic information processing technologies to monitor patient status at a distance". It can also be defined as the use of information technology to monitor patients who are not located in the same place of the health care provider. Meystre reported how the telemonitoring systems have been successful adopted in cardiovascular, hematologic, respiratory, neurologic, metabolic, and urologic domains [22]. In fact, it may help people stay healthy, and in their homes, longer [8]. Better follow-up of patients is a convenient way for patients to avoid travel and to perform some of the more basic work of healthcare for themselves, thus reducing the corresponding overall costs [1, 36]. In addition to objective information coming from technological monitoring, most systems include also subjective data based on asking to the user a sort of questionnaires to the patient's about health and comfort [21]. This questioning can take place automatically over the phone or in other ways, as on-line formats.

Home sensor technology may create a new opportunity to reduce costs. An interest has therefore emerged in using home sensors for health promotion [16]. One way to do this is by telemonitoring and home support systems (TMHSSs), which are aimed at remotely monitoring patients who are not located in the same place of the health care provider.

Relying on a combination of subjective and objective information similar to what would be revealed during an on-site appointment, therapists and caregiver may make decisions about the patient's treatment, or improve it. Moreover, all of this information is centralized in the telemonitoring platform, which can exploit the information to help therapists and care givers to take better decisions.

2.3   Teleassistance

A definition of teleassistance has been given in [5]: "remote, automatic and passive monitoring of changes in an individual's condition or lifestyle (including emergencies) in order to manage the risks of independent living". Thanks to Teleassistance, we can offer the possibility to the users to be connected with therapists and caregivers as well as relatives and family, allowing people with special needs to be independent. One example of this kind of systems can be the one developed in the SAAPHO project [31], which was aimed to offer active and independent living for elderly, based on a teleassistance system.

Teleassistance systems should process and analyse all this information in order to detect abnormal situations or improve the quality of life of the users helping them to carry out daily activities. This information can be processed at the user side in order to offer the possibility of self-management and should be also available to therapists and caregivers allowing them to offer assistance in any situation.

3   A Generic Platform for Tele*

To provide Tele* solutions for, in principle, whatever kind of end-users, as well as to give a continuous and efficient support to therapists and caregivers, we defined and developed a generic platform. The aim of the proposed solution is to be as flexible as possible in order to be able to provide telerehabilitation, telemonitoring, teleassistance or a conjunction of them, depending on the real situation.

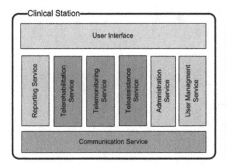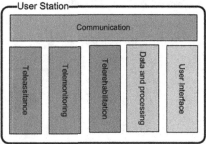

Fig. 1 Main components of the generic Tele* platform: Clinical Station (on the *left*) and User Station (on the *right*)

As stated above, the main actors of Tele* are, from one side, the end-users (normally patients or elder people) and, on the other side, therapists and caregivers. Thus, our generic platform is composed of two main modules, called "stations": Clinical Station and User Station (see Fig. 1). Both stations have been defined as standalone solutions. In so doing, the Clinical Station offers the proper functionality to the professionals independently of the User Station implementation. Similarly, the User Station offers its functionality to the patient without taking into account the implementation of the Clinical Station.

3.1 The Clinical Station

The Clinical Station is the entry point of therapists and caregivers (see Fig. 1). Its main functionalities are: Reporting, Therapy Management System, Monitoring, Teleassistance, Administration Services, User Management, as well as the User Interface and the Communication Service.

The *Reporting Service* module is aimed at giving all the information regarding the processed data. Which information and how to display it depend on the real implementation (from graphs to complete reports). As an example, let us consider the general report given in the ActivApp Clinical Station depicted in Fig. 2.

Through the *Telerehabilitation Service* module, therapists may plan, schedule, access to the results, and personalize the prescription of rehabilitation tasks inside her/his therapeutic range (i.e. motivating and supporting her/his progress), in order to get the better therapeutic results. The sessions can be configured, setting the type of tasks that the user will execute, their order as well as the difficulty level, and specific parameters depending on the exercise. Additionally, the Clinical Station allows therapists to establish an occurrence pattern for the session along the time. If the same session must be executed several times, they can set the type of occurrence and its pattern to make the session occur at programmed times in the future. Upon completion of the game session execution on User Station, results are sent back to the

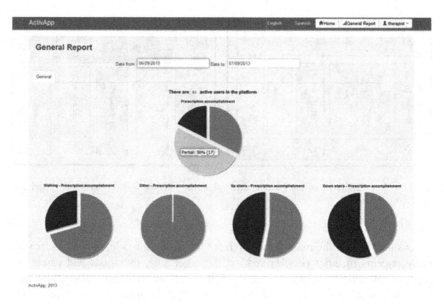

Fig. 2 Clinical Station: General report in ActivApp

Clinical Station for review by therapists that are notified with an alert in the dashboard indicating that the user has completed the session. The Clinical Station provides a session results view and an overview of completed sessions to map progress, which shows session parameters and statistics along the specific results for each game in the session (see Fig. 3 with an example from BackHome). Moreover, aggregated game results for a specific user are also provided. This provides statistic results on all the executed occurrences of a given game by a user, whether included in a session or executed as a standalone game, so professionals can have a better insight on the user progression in a specific area. The end-user status as well as her/his rehabilitation activities are continuously monitored by the *Telemonitoring Service* module that relies on suitable sensors: wearable, physiological, biometric sensors, environmental sensors, SmartHome devices, devices that allow interaction activities or devices to perform rehabilitation tasks (e.g. a robot). Specifically, the wearable sensors enable the monitoring of fatigue, spasticity, stress, and further user's conditions. Additionally, environmental sensors are used to monitor temperature, humidity, movement (motion sensors) and the physical position of the user (location sensors). SmartHome devices aim to enhance the users physical autonomy by supporting the user to undertake heir daily life activities. Internet-connected devices enable social autonomy of users helping her/him to keep in touch with relatives and friends. The actual set of devices depend on the real implementation, for instance ActivApp relies only on a wearable sensor (i.e. the end-user's smartphone) whereas in BackHome domotic and environmental sensors and actuators have been used. Moreover, personalised information can be captured through a combination data coming from the sensor-based system and of questionnaires. This information will be fused with that gathered when

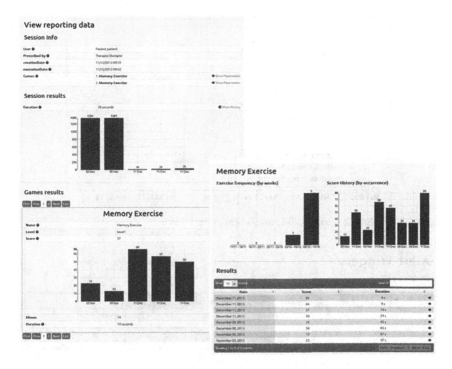

Fig. 3 Clinical Station: rehabilitation session and game results in BackHome

the individual is interacting with the system. The data will be used to inform the system of the users' behavior, social autonomy, rehabilitation or to other support tasks. For instance, in BackHome the professional staff can see the inferred data about the quality of life of the user (see in Fig. 4).

The *Teleassistance Service* module gives information about habits, behaviors, abnormal situations and emergencies. Several kinds of techniques (from rule-based to advanced IA algorithms) can be used to detect the current status or emergencies, depending on the real application. Figure 5 shows an example of rule definition aimed at detecting the status of the end-user relying on the raw information coming from the adopted sensor-based system. In the example in the Fig. 5, the therapist sets a rule to raise an alarm if the user spends more than ten hours in bed in a day. Thus, the system controls the sleep pattern of the user to detect abnormal situations. In doing so, if the user has been in bed for more than 10 h and the user station detects it, an alert is raised to the therapist who receives a message at the dashboard of the system showing the alert that has been triggered and the user who is involved.

The *Administration Manager* module is aimed at providing administrative functionality at the high level (i.e. therapists and end-users creation, as well as definition of rehabilitation games or users' questionnaires). Therapists and caregivers do not interact directly with this module, being the system administrator the unique user

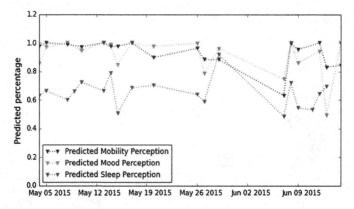

Fig. 4 Clinical Station: quality of life assessment in BackHome

Alert Triggers

+ New trigger

Show 10 ▼ entries   Search:

Name ▲	Occurrence ⬍	Creation Date ⬍	⬍
Inactivity	Daily	12 February, 2014	✕

Showing 1 to 1 of 1 entries   First Previous 1 Next Last

New Trigger

┌─ Basic information ───┐

              Name   Trigger   Occurrence   Continuous ▼

└──┘

┌─ Conditions ──┐

  ⤷ Logic Condition   Operations   Operand ▼
    ⤷ AND
      ├ Room equals Bedroom   Variable   hours in bed ▼
      └ Hours in bed greater than 10
                                                           Operation   lower than ▼

                                                           Value   3 ▼

└──┘

┌─ Actions ───┐

  Alert -> Send Alert to Therapist   Device target   Alert ▼

                                                           Action performed Send Alert to all ▼

                                                                                        Add

  Remove

└──┘

Fig. 5 Clinical Station: rule definition in BackHome

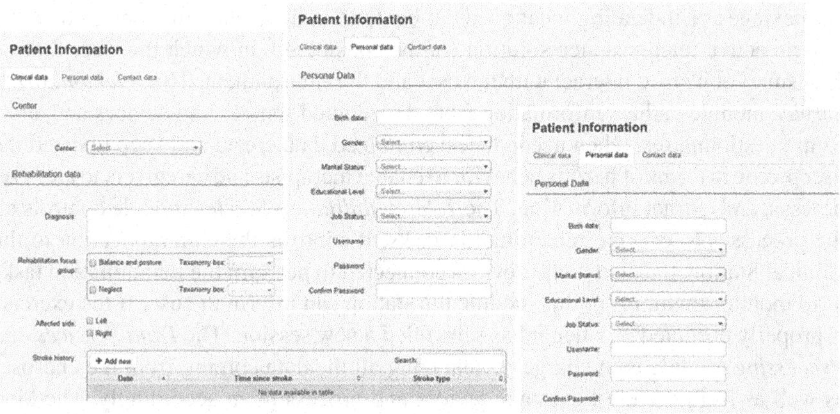

Fig. 6 Clinical Station: general information for a patient in Rewire

in charge of performing those actions. On the contrary, therapists and caregivers, through the *User Management* service, may create new end-users in the platform and manage their information (from personal to therapy-based). Some data from the user that manages this service can be personal data (e.g. name, date of birth, marital status), contact data (e.g. phone number, address) and clinical data (e.g. diagnosis, health center, rehabilitation therapies, treatments, etc.). Figure 6 shows an example of user information provided by the Clinical Station. The *Communication Service* module is the responsible for connecting the different information sources (from hospitals to end-user devices and apps), and also to communicate with the User Station. This communication will depend on the technical configuration of both stations in the real scenario. It could be implemented as a REST service in the case of the User Station being implemented as a mobile app (as in the case of ActivApp, as described later) or as a specific protocol in the case of using external devices, as, for instance, a Kinect (as in the case of REWIRE, as described later). The *User Interface* is a Web-based interface that allows interaction through any available device (e.g. a PC, a tablet, a smartphone) and from any place connected to the Internet.

3.2 The User Station

User Station is the end-user's entry point to the system. It may have different incarnations depending on the real application and end-users' needs (see Fig. 1).

The *Communication Service* module is the one in charge of communicating with the Clinical Station. The corresponding protocol, as already said, depends on the real application. The *Teleassistence Service* module determines what action has to be done to help the user to complete daily activities or to solve emergency problems. It may adopt a passive solution, in which the User Station can interact with the user only by messages or indicating what needs to be done in different situations. Moreover, also an active teleassistance solution can be considered, in which the User Station uses smart objects to interact with the user and the environment. The *Telemonitoring Service* module gathers information from the adopted sensors and devices and, also, from questionnaires, when needed. The combined data create and keep updated the user profile in terms of her/his behavior to enable therapists and caregivers to analyze, process, and extract information. The *Telerehabilitation Service* module controls all the processes to execute rehabilitation tasks. It controls the communication to the Clinical Station and also to the devices connected to perform the rehabilitation tasks or to monitor them. From this module the station can inform the user if the exercise is properly executed or when it has scheduled a new session. The *Data Services and Processing* module is in charge of analyzing all the data coming from the end-user as well as from the used external devices and process them accordingly. The kind of analysis and process depends on the real application. Basically, this module will provide recommendations and continuous support to the end-user as well as suitable information, alarms, alerts and notifications to therapists and caregivers. More details will be given for each real application in the following sections. The *User Interface* may be implemented as a mobile application (as in ActivApp) or may run on a computer connected to a TV (as in REWIRE) or with more advanced and complex devices (as in BackHome).

4   Real Uses Cases of the Tele* Platform

The generic Tele* platform proposed in this chapter has been customized to be used in three real end-user scenarios, under the umbrella of three research projects: REWIRE, ActivApp, and BackHome.

4.1   Telerehabilitation at Home: The REWIRE Project

In the European R + D project REWIRE,[1] we customized our generic Tele* platform to be used for motor rehabilitation of patients at home. The main aim of REWIRE is to provide the possibility of continuing the rehabilitation at patient's home under remote monitoring by the hospital. The project has developed, integrated and tested a virtual reality system based on a multi-level rehabilitation platform. This platform, deployed at the patients' homes, enables home-based effective rehabilitation to improve disabilities and functions [4].

---

[1]http://www.rewire-project.eu/.

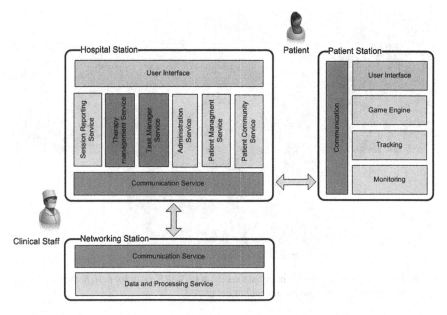

Fig. 7 The implementation of the generic Tele* platform in REWIRE

4.1.1 The Actual Implementation of the Generic Tele* Platform

In REWIRE, the generic Tele* platform has been implemented by relying on three stations, as shown in Fig. 7: Patient Station, Hospital Station, and Networking Station.

Patient Station The Patient Station is used by the patient and her/his caregivers at home to perform supervised rehabilitative training inside a virtual environment. The patient is prompted by the *User Inteface*, implemented as a virtual environment, to follow the exercises, planned by the clinicians at the hospital, through adequate simple games (see Fig. 8) [3]. Patient's motion and interaction data with the virtual environment is saved while patient's daily life is monitored by a set of wearable sensors. All this gathered information is used to better tune the rehabilitation sessions and set-up a proper level of challenge for the patient, assess potential risks and advice clinicians on the therapy. The *Game Engine* module is responsible of providing games for rehabilitation. This module translates the information from the Hospital Station to schedule the games and adapt them to the requirements indicated by the therapist. The *Tracking* module is in charge of measuring the motion of the patient. The *Monitoring* module gathers information from the sensors used to track the exercises, physiological sensors and ambient sensors. These data are processed to provide an active feedback to both the patient and the clinicians, to tune the rehabilitation session and to provide data for risk evaluation [26].

Hospital Station The Hospital Station is used by the therapists at the hospital. Working as a web application, it helps therapists to define, tune and monitor the exercises of the rehabilitation program at home, taking into account all the activity data

Fig. 8 Screenshots of the some games implemented in REWIRE, from *left* to *right* the fruit catcher game, the hay collect game and the animal hurdler game

collected by the Patient Station during the exercises. It implements the functionalities described in Sect. 3.1.

Networking Station The Networking Station is in charge of analyzing the data, i.e. to implement the *Data Service and Processing* module that is part of the User Station in the generic Tele* platform. This station analyses the multi-parametric data that come from the individual patient day by day rehabilitation, to compare and interpret the results on different patient populations. Data mining procedures have been developed to discover features and trends of rehabilitation treatments in the different hospitals involved in the project [14].

Communication Protocol The Patient Station, installed at patient's home, communicates with the Hospital Station trough web services. These web services define the functionalities that the Patient Station can do, such as reporting the results of the rehabilitation exercises. All the communication is secure under the HTTPS protocol. The Networking Station is a standalone application which imports the data from the Hospital Station, directly.

4.1.2   Results

The REWIRE system has been installed in 2 health institutions, one located in Spain and one in Switzerland. In total, 10 patients have participated in the trials (six with a diagnosis of stroke and two suffering neglect). After 2 weeks of training in the hospital, four of the six patients with a stroke diagnosis started a 3 month rehabilitation process. The two patients with a diagnosis of neglect had 1 month rehabilitation process. Healthcare professionals scheduled 347 sessions in total during the 3 months length of the study, devoting an average time of 17.33 ± 2.07 min to schedule five sessions per week. Patients underwent 80 % of scheduled sessions, with an average of 46.17 ± 13.98 sessions per patient. During the 2 weeks of hospital training for patients with a diagnosis of neglect, 12 sessions were registered in total. Patients suffering neglect received an average of 6.00 ± 2.83 sessions. The average time spent in each session was 25.08 ± 7.69 min. For further details of the results, the interested reader may refer to [4].

4.2 Telerehabilitarion and Telemonitoring Outdoors: The ActivApp Project

In the ActivApp project, funded by ACC1ó (Generalitat de Catalunya),[2] we customized the generic Tele* platform to provide mHealth tools to personalize the health monitoring and treatment through mobile applications, particularly for patients suffering COPD. ActivApp is focused on remote monitoring of patients with COPD. By using a mobile application and the accelerometer embedded in the smartphone, in conjunction with a web platform for therapists, end-users physical activity is remotely prescribed and monitored, in a non-obtrusive and continuous way. Applying gamification techniques and personalized user interfaces, the adherence to the acquired good habits is promoted.

4.2.1 The Actual Implementation of the Generic Tele* Platform

Figure 9 shows how the generic Tele* platform has been customized in ActivApp. It is composed of a mobile app for the patient (Patient Station) together with a Web-based application for therapists (Therapist Station).

Patient Station The Patient Station is a mobile app composed of five main modules together. Additionally, the *Communication Service* is in charge of performing the communication with the Therapist Station. The *User Interface* prompts the patient to carry out a certain prescribed type and intensity of physical activity and to answer short questionnaires and scales, automatically monitors the treatment through data collected from the smartphone sensors and processed with stream data mining algorithms and motivates the user to exercise. The *Motivation* module is responsible of engaging the patient in the application according to persuasive computing techniques. In particular, gamification techniques have been used to motivate the patient and to keep her/him in action. The module uses the information processed about activity to reward the user with badges. The user can get this badges only if enough activity has been performed—i.e. virtual routes have been completed. These routes are divided in sections which can be completed in several days, once the user com-

---

[2]http://accio.gencat.cat/.

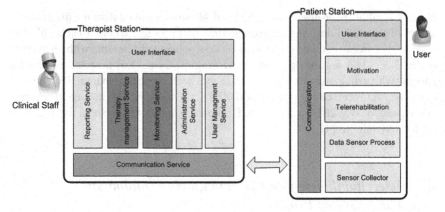

Fig. 9 The implementation of the generic Tele* platform in ActivApp

pletes all the sections wins a badge. The routes are near the user's living place to be easily recognizable by the user. Moreover, the patient is also asked to complete simple questionnaires, planned by the therapists. Currently, the COPD Assessment Test) [18] and the Borg [2] questionnaires have been implemented. Figure 10 shows some of those functionalities, from left to right: main screen with the statistics, Borg questionnaire and motivational routes. The *Telerehabilitation* module is in charge of following up the rehabilitation activities assigned by clinicians and performed by the users. The *Data Sensor Process* module, based in a Hoeffding Tree [15] with Naive Bayes Adaptative leaf prediction and built over Android, is aimed at processing all the gathered data. The system create models off-line and then they are included in the app. Thanks to these models the app can detect no activity, walking, cycling, going up and down stairs. The *Sensor Collector* is in charge of gathering all the data by the smartphone.

Therapist Station Through the Therapist Station, therapists may schedule fitness activity and plans to the patient. These include the required level of activity, the duration, and put some goals to be reached day by day. To help the user to be motivated, the system shows different emoticons. These emoticons indicate the level of the plan reached every day divided in three levels: more than 75 % of compliance, between 50 and 75 % and less of 50 % of compliance.

Communication Protocol The two stations share information using REST messages and push notifications under HTTPS protocol in order to be secure. The structure and order of the messages are defined by a custom specification.

4.2.2 Results

The system has been tested with patients with COPD to determine the feasibility to use this kind of platforms in real situations. The field trials are focussed in the usability and validation of the system rather than on clinical results. The evaluation was done

Fig. 10 Screen shots of the patient station of ActivApp

by a group of 6 users (6 male, M = 68 ± 5 years). They have used the application during 8 weeks to monitor their activity and to follow the indications of the therapists. Usability results shows good results: the 65 % of the users evaluate the system a good and useful application; the 18 % just as normal application; and the 17 % as a bad application. The users highlight the easy to use of the main functionalities and the usefulness of the application. The possibility to offer more information to the clinical staff is really well evaluated.

With the data collected in these trials, the activity classification model has been tested. On one hand, the results are over the 90 % in accuracy: 93.7 % in case of walking and 95.7 % in case of no activity. On the other hand, the stairs activity is 63.9 % to go down stairs and 66.1 % in case of the detection of going upstairs. Riding a bicycle has an accuracy of the 88.7 %.

4.3 A Complete Tele* System: Telerehabilitation, Telemonitoring, and Teleassistance at Home in the BackHome Project

In the BackHome project,[3] the generic Tele* platform has been customized to provide telerehabilitation, telemonitoring, and teleassistance to people with severe disabilities that go back to home after a discharge.

---

[3] http://www.backhome-fp7.eu/.

BackHome aims to study the transition from the hospital to the home, focusing on how people use Brain Computer Interfaces (BCIs) in both settings [23]. Moreover, it is aimed to learn how different BCIs and other assistive technologies work together and can help clinicians, disabled people, and their families in the transition from the hospital to the home. The final goal of BackHome is to reduce the cost and hassle of the transition from the hospital to the home by developing improved products. To produce applied results, BackHome will provide: new and better integrated practical electrodes; friendlier and more flexible BCI software; and better telemonitoring and home support tools [10].

4.3.1 The Actual Implementation of the Generic Tele* Platform

The BackHome system implements the generic Tele* platform through two stations, the User Station and the Therapist Station, as depicted in Fig. 11.

User Station The User Station, composed of six modules, is completely integrated into the end-user's home. The *BCI Interface* allows the end user to interact with the BCI components (a screen user interface together with a BCI block and BCI equipment). With that interface, the end-user may interacts with all the provides services via control matrices [11]. The *Cognitive Rehabilitation* module provides three rehabilitation tasks in form of serious games [34]: memory-cards; find a category;

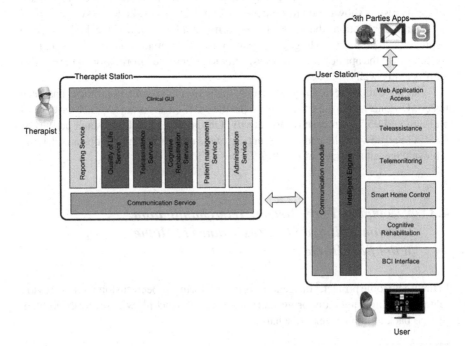

Fig. 11 The implementation of the generic Tele* platform in BackHome

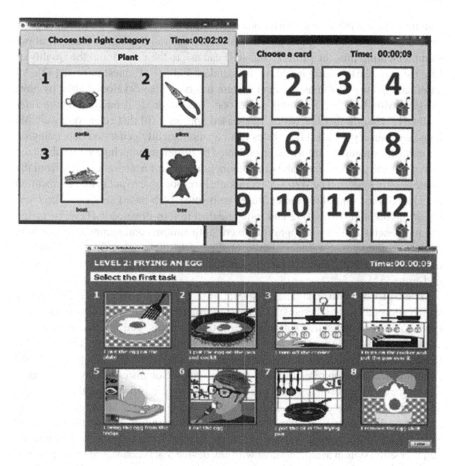

Fig. 12 Screenshots of the Cognitive Rehabilitation tasks in BackHome

and daily-life activities, which allow users to develop skills that can be applied to real life tasks. Figure 12 shows one screenshot for each implemented serious game. The *Smart Home Control* module is in charge of handling all the smart home devices installed at the user's home. In so doing, it provides control over the built environment and free standing electrical goods allowing the user to control them through the BCI. The system integrates ON/OFF switches and power meters connected with appliances (i.e. a light, a fan, and a radio). The *Telemonitoring* module has been implemented through a sensor-based telemonitoring and home support solution [24, 27]. The sensor-based system is able to monitor indoor and outdoor activities by relying on a set of z-wave sensors. They send the retrieved data to a collector (based on Raspberry pi), which collects all the retrieved data and securely redirects them to the Cloud where they will be stored, processed, mined, and analyzed. As for outdoor activities, we use the user's smartphone as a sensor by relying on Moves,[4] an app for smartphones able to recognize physical activities and movements by transportation. Complementing this, the professional staff can ask to the users about the Quality of Life using a questionnaire based on the standard EQ-5D-5L questionnaire [33]. An automatic quality of life assessment system has been also developed able, by now, to assess Mobility [23], Sleeping, and Mood. Teleassistance is handled by the *Teleassistance Module* module with a rule-based approach [6] that relies on a suitable language, namely ATML [12]. Rules can be automatically generated depending on the context by using the information collected by the telemonitoring system about the activity of the user and his habits in conjunction with information coming from the home control system. The *Web Access* module is responsible to put in communication the User Station and, thus, the end-user with some Web-based services. The *Intelligent Engine* aims to analyze and mine the data in four-dimensions [29]: detection of emergencies, activity recognition, event notifications, and summary extraction. Moreover, a context-awareness based quality of life telemonitoring methodology has been defined and implemented to automatically assess the quality of life of the end-users [35].

Therapist Station The Therapist Station is a Web-based easy-to-use service which allows a remote therapist to access information stored in the Cloud and gathered from users and sensors around them: users' inputs, activities, selections and sensor data. This information takes the form of system usage reports, rehabilitation tasks results and quality of life assessment, and supports those therapists to make informed decisions on rehabilitation planning and personalisation as well as remote assistance and support action triggering defining rules in order to be alerted in case some anomalies are detected. The scalable and robust cloud storage of data and ubiquitous Web access provides the needed flexibility in order to get the maximum potential out of the telemonitoring and home support features because the therapist can access the station at any moment with any device that is connected to the Internet.

Communication Protocol User Station and Therapist Station establish a bidirectional and secure communication over https where SOAP messages are used to send all the data related to the user. This approach allows the definition of a communication protocol using message schemas (XSD) that allow easy message validation, implemented at both ends of the communication. The communication protocol is based on authenticated user sessions, forcing the user station to first authenticate the user. Once authenticated, the therapist station acknowledges the communication by generating a unique token that identifies a user session. This token is then attached to every message shared between user station and therapist station until they close the communication.

---

[4]http://www.moves-app.com/.

4.3.2   Results

The overall system ran for 6 weeks at three end-user' homes in Belfast and was also tested in some homes in Würzburg. Before installing the system in end-user's homes, we tested it on a 40 years-old able-bodied user's home in Barcelona [28].

For each user the following information was given: when the user was at home, away, or received visits day by day and a summary of the different locations in which the user was during a period of 1 month. Collected data have also been used to automatically assess quality of life of monitored users in terms of Mobility, Sleeping, and Mood; results are reported to therapists for study and follow up.

Participants were enthusiastic about their experience evaluating the BackHome prototype in their own home. Ultimately there were challenges; however the learning from this evaluation is essential to realise the fundamental goal of moving BCI into peoples' homes as an AT to support independent living. Home users were able to complete 61 and 72 % of tasks set for them over the six weeks. Satisfaction with the system was strongly linked to the systems responsiveness throughout the evaluation on the BCI satisfaction scale. Additionally, both of the home users were satisfied with the BCI on the eQUEST 2.0.

5   Conclusions

Telerehabilitation, telemonitoring, and teleassistance (Tele* for short) are cost-effective solutions that help therapists and caregivers, as well as end-users, to follow a given therapy remotely. Very important is the role of therapists and caregivers that may access relevant information about their patients and assisted people. In this chapter, we proposed and described a generic Tele* platform aimed at supporting all those actors. In principle, the proposed platform may be used in any Tele* real scenario. To show its effectiveness, we illustrated how it has been customized in three real cases: a telerehabilitation system for people with motor impairments; a mobile telerehabilitation and telemonitoring system for people affected by COPD; and a complete telerehabilitation, telemonitoring, and teleassistance system for people with severe disabilities that go back to home after a discharge.

Acknowledgments The research leading to these results has received funding from the European Community's, Seventh Framework Programme FP7/2007-2013, BackHome project Grant Agreement No. 288566 and REWIRE project Grant 287713.

References

1. Artinian, N.: Effects of home telemonitoring and community-based monitoring on blood pressure control in urban African Americans: a pilot study. Heart Lung **30**, 191–199 (2001)
2. Borg, G.: Borg's perceived exertion and pain scales. Hum. Kinet. (1998)

3. Borghese, N., Lanzi, P., Mainetti, R., Pirovano, M.: Apparatus and method for rehabilitation employing a game engine. In: US Application, US Application (2013)
4. Borghese, N.A., Murray, D., Paraschiv-Ionescu, A., de Bruin, E.D., Bulgheroni, M., Steblin, A., Luft, A., Parra, C.: Rehabilitation at home: a comprehensive technological approach. In: Virtual, Augmented Reality and Serious Games for Healthcare 1, pp. 289–319. Springer (2014)
5. Bower, P., Cartwright, M., Hirani, S.P., Barlow, J., Hendy, J., Knapp, M., Henderson, C., Rogers, A., Sanders, C., Bardsley, M., et al.: A comprehensive evaluation of the impact of telemonitoring in patients with long-term conditions and social care needs: protocol for the whole systems demonstrator cluster randomised trial. BMC Health Serv. Res. 11(1), 184 (2011)
6. Casals, E., Cordero, J.A., Dauwalder, S., Fernández, J.M., Solà, M., Vargiu, E., Miralles, F.: Ambient intelligence by atml: rules in backhome. In: Lai, C., Giuliani, A., Semeraro, G. (eds.) Emerging Ideas on Information Filtering and Retrieval. DART 2013: Revised and Invited Papers (2014)
7. Chen, S., Thomas, J., Glueckauf, R., Bracy, O.: The effectiveness of computer-assisted cognitive rehabilitation for persons with traumatic brain injury. Brain Inj. 11(3), 197–210 (1997)
8. Cordisco, M., Benjaminovitz, A., Hammond, K., Mancini, D.: Use of telemonitoring to decrease the rate of hospitalization in patients with severe congestive heart failure. Am. J. Cardiol. 84(7), 860–862 (1999)
9. Dou, Z., Man, D., Ou, H., Zheng, J., Tam, S.: Computerized errorless learning-based memory rehabilitation for chinese patients with brain injury: a preliminary quasi-experimental clinical design study. Brain Inj. 20(3), 219–225 (2006)
10. Edlinger, G., Hintermller, C., Halder, S., Vargiu, E., Miralles, F., Lowish, H., Anderson, N., Martin, S., Daly, J.: Brain neural computer interface for everyday home usage. In: HCI International 2015 (2015)
11. Farwell, L.A., Donchin, E.: Talking off the top of your head: toward a mental prosthesis utilizing event-related brain potentials. Electroencephalogr. Clin. Neurophysiol. 70(6), 510–523 (1988)
12. Fernández, J.M., Torrellas, S., Dauwalder, S., Solà, M., Vargui, E., Miralles, F.: Ambient-intelligence trigger markup language: a new approach to ambient intelligence rule definition. In: 13th Conference of the Italian Association for Artificial Intelligence (AI*IA 2013). CEUR Workshop Proceedings, vol. 1109 (2013)
13. Field, M.J. (ed), Committee on Evaluating Clinical Applications of Telemedicine, I.o.M.: Telemedicine: A Guide to Assessing Telecommunications for Health Care. The National Academies Press (1996)
14. Globas, C., Ženko, B., Džeroski, S., Luft, A.: Predicting disability and quality of life after ischemic stroke. In: Spring Workshop on Mining and Learning Computational Intelligence and Games (CIG), 2012 (2012)
15. Hoeglinger, S., Pears, R.: Use of hoeffding trees in concept based data stream mining. In: Third International Conference on Information and Automation for Sustainability, 2007. ICIAFS 2007, pp. 57–62. IEEE (2007)
16. Intille, S.S., Kaushik, P., Rockinson, R.: Deploying context-aware health technology at home: human-centric challenges. Human-Centric Interfaces for Ambient Intelligence (2009)
17. Johansson, B., Tornmalm, M.: Working memory training for patients with acquired brain injury: effects in daily life. Scand. J. Occup. Ther. 19(2), 176–183 (2012)
18. Jones, P., Harding, G., Berry, P., Wiklund, I., Chen, W., Leidy, N.K.: Development and first validation of the copd assessment test. Eur. Respir. J. 34(3), 648–654 (2009)
19. Kang, S.H., Kim, D.K., Seo, K.M., Choi, K.N., Yoo, J.Y., Sung, S.Y., Park, H.J.: A computerized visual perception rehabilitation programme with interactive computer interface using motion tracking technology—a randomized controlled, single-blinded, pilot clinical trial study. Clin. Rehab. (2009)
20. Lundqvist, A., Grundström, K., Samuelsson, K., Rönnberg, J.: Computerized training of working memory in a group of patients suffering from acquired brain injury. Brain Inj. 24(10), 1173–1183 (2010)
21. Martń-Lesende, I., Orruño, E., Cairo, C., Bilbao, A., Asua, J., Romo, M., Vergara, I., Bayn, J., Abad, R., Reviriego, E., Larrañaga, J.: Assessment of a primary care-based telemonitoring

intervention for home care patients with heart failure and chronic lung disease. The TELBIL study. BMC Health Serv. Res. **11**(56) (2011)

22. Meystre, S.: The current state of telemonitoring: a comment on the literature. Telemed. J. E Health **11**(1), 63–69 (2005)

23. Miralles, F., Vargiu, E., Casals, E., Cordero, J., Dauwalder, S.: Today, how was your ability to move about? In: 3rd International Workshop on Artificial Intelligence and Assistive Medicine, ECAI 2014 (2014)

24. Miralles, F., Vargiu, E., Dauwalder, S., Solà, M., Fernández, J., Casals, E., Cordero, J.: Telemonitoring and home support in backhome. In: Proceedings of the 8th International Workshop on Information Filtering and Retrieval co-located with XIII AI*IA Symposium on Artificial Intelligence (AI*IA 2014) (2014)

25. Patel, S., Park, H., Bonato, P., Chan, L., Rodgers, M.: A review of wearable sensors and systems with application in rehabilitation. J. Neuroeng. Rehab. **9**(1), 21 (2012)

26. Pirovano, M., Mainetti, R., Baud-Bovy, G., Lanzi, P.L., Borghese, N.A.: Self-adaptive games for rehabilitation at home. In: 2012 IEEE Conference on Computational Intelligence and Games (CIG), pp. 179–186. IEEE (2012)

27. Rafael-Palou, X., Vargiu, E., Dauwalder, S., Miralles, F.: Monitoring and supporting people that need assistance: the backhome experience. In: Lai, C., Giuliani, A., Semeraro, G. (eds.) DART 2014: Revised and Invited Papers (in press)

28. Rafael-Palou, X., Vargiu, E., Miralles, F.: Monitoring people that need assistance through a sensor-based system: evaluation and first results. In: Proceedings of AI-AM/NetMed 2015 Artificial Intelligence and Assistive Medicine. Proceedings of the 4th International Workshop on Artificial Intelligence and Assistive Medicine co-located with the 15th Conference on Artificial Intelligence in Medicine (AIME 2015), CEUR Workshop Proceedings, vol. 1389 (2015)

29. Rafael-Palou, X., Vargiu, E., Serra, G., Miralles, F.: Improving activity monitoring through a hierarchical approach. In: ICT for Ageing Well Conference (2015)

30. Rego, P., Moreira, P.M., Reis, L.P.: Serious games for rehabilitation: a survey and a classification towards a taxonomy. In: 2010 5th Iberian Conference on Information Systems and Technologies (CISTI), pp. 1–6. IEEE (2010)

31. Rivero-Espinosa, J., Iglesias-Pérez, A., Gutiérrez-Dueñas, J.A., Rafael-Palou, X.: Saapho: an AAL architecture to provide accessible and usable active aging services for the elderly. ACM SIGACCESS Access. Comput. **107**, 17–24 (2013)

32. Tam, S.F., Man, W.K.: Evaluating computer-assisted memory retraining programmes for people with post-head injury amnesia. Brain Inj. **18**(5), 461–470 (2004)

33. The Euroqol Group: Euroqol a facility for the measurement of health-related quality of life. Health Policy **16**, 199–208 (1990)

34. Vargiu, E., Dauwalder, S., Daly, J., Armstrong, E., Martin, S., Miralles, F.: Cognitive rehabilitation through bnci: serious games in backhome. In: Mller-Putz, G.R., Bauernfeind, G., Brunner, C., Steyrl, D., Wriessnegger, S., Scherer, R. (eds.), Proceedings of the 6th International Brain–Computer Interface Conference, pp. 36–39 (2014)

35. Vargiu, E., Fernández, J.M., Miralles, F.: Context-aware based quality of life telemonitoring. In: Lai, C., Giuliani, A., Semeraro, G. (eds.) Distributed Systems and Applications of Information Filtering and Retrieval. DART 2012: Revised and Invited Papers (2014)

36. Vincent, J., Cavitt, D., Karpawich, P.: Diagnostic and cost effectiveness of telemonitoring the pediatric pacemaker patient. Pediatr. Cardiol. **18**(2), 86–90 (1997)

37. Zickefoose, S., Hux, K., Brown, J., Wulf, K.: Let the games begin: a preliminary study using attention process training-3 and lumosity? brain games to remediate attention deficits following traumatic brain injury. Brain Inj. **27**(6), 707–716 (2013)

Printed in the United States
By Bookmasters